Robot
Destruction

Peter Bennett

Peter Bennett

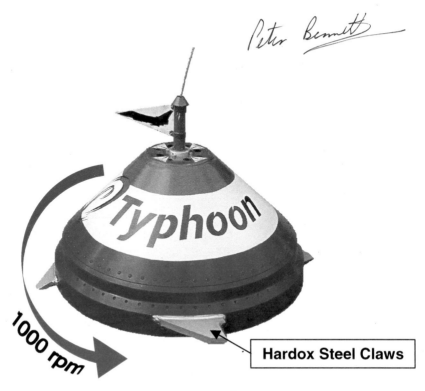

Hardox Steel Claws

1000 rpm

Typhoon 2

First Published in 2004
by
P J Bennett
15 Braepark Road
Edinburgh EH4 6DN
Scotland

ISBN – 0-9547388-0-2

Printed in Scotland by CS&S Publishing Group

****

Proceeds from the sale of this book go to the Edinburgh Air Cadets Typhoon Robots Project

Introduction

The Edinburgh Air Cadets and their 'Typhoon' robots have been winning the honours in "Robot Wars" and "Technogames" since 2001. There are fighting robots in four weight categories and even a robot bicycle (Byphoon). In 2004 their unseeded heavyweight robot 'Typhoon 2' became the **UK Grand Champion**.

This book describes the Typhoon Robots Project from its inception and shows how simple mathematics and physics is used to optimise the robot designs and make them truly formidable machines. Robotics is a subject which excites young people and shows them that science and engineering can be real fun.

The book has not been written as a school textbook and it is not focused on any exam syllabus. Instead it aims to show how many of the maths and physics concepts that pupils encounter in their theoretical studies at school have been used by their contemporaries to create a world class machine that has beaten the country's best robots.

Mathematics is included in three of the ten chapters to explain the compromises that must be considered to achieve a winning design. The majority of the concepts are covered by students at Standard Grade/GCSE and the practical significance of each calculation is explained.

I am indebted to my colleagues from BAE Systems, who have given up their spare time to assist the cadets with this project. We have all been amazed at the dedication and enthusiasm of these young people. They have fully embraced the scientific concepts and enthusiastically regale their friends with the theory behind the Typhoon design and the significance of the major design parameters.

Peter Bennett

**870 (Dreghorn) Squadron Robot Teams With Six of Their
Robot Wars and Technogames Robots**

Foreword

Professor John Roulston OBE
Technical Director, BAE SYSTEMS Avionics

I earned the privilege of writing this foreword simply by suggesting to Peter Bennett that the Typhoon Robot story merited a book. Immediately, Peter wrote a book, this book. It is not the book I imagined, though it makes some concession to the vision I tried to impart. I saw the Typhoon project as the best ever answer to two questions that kept arising, one from school-children, "what good is mathematics?" and the other from teachers, "how do I get realistic and exciting teaching examples for my mathematics class?" Hence, I had in mind an illustrated text-book, aligned to the Scottish syllabus. Something to kindle enthusiasm in applied mathematics and encourage pupils to take up a hard option that has been receding in popularity in recent years. Peter started off true to the vision and the compelling narrative of Typhoon's creation took over. This book is narrative, an amazing narrative of the efforts of the Air Cadets and the engineers who supported them through invention, construction, test, refinement and ultimately astonishing success. Maybe success is not so astonishing. It was calculated at each step. The Cadets were organised into an effective multi-disciplinary team. Plans were made, tasks were managed, sub-systems were tested, problems were faced and solved and the whole Typhoon concept came together, step by anxious step, as the thrill of engineering opened up a new world of experience. Peter's book imparts a lot of this. It also carries the data that allows quantitative dissection of the design, so that teachers might be able to formulate their examples. Certainly, it's a work of intellectual generosity and it can't help but inspire, so in its way it fulfils the educational vision. After reading this there can be no doubt of the value of projects like Typhoon. We don't do enough of them. If we did, our society would increase in technological sophistication and we would be the more prosperous for it. Now the Typhoon team has the challenge of beating their own creation. Good luck and happy roboteering.

Edinburgh,
March 2004

Contents

Foreword		4
Chapter 1	Take Spare Boxer Shorts to Watch Robot Wars	7
Chapter 2	First Combat	18
Chapter 3	Technogames	22
Chapter 4	Off To War Again	26
Chapter 5	Weaknesses Must Be Minimised	37
Chapter 6	Depleted Uranium, Titanium or Unobtainium	40
Chapter 7	Engaging and Avoiding the Enemy	45
Chapter 8	A Weapon of Mass Destruction	58
Chapter 9	We Need a Lightweight Nuclear Power Source	75
Chapter 10	How Can We Tame This Mechanical Beast	84
Appendix A	Our Sponsors	97
Appendix B	The Typhoon Family	98
Appendix C	Design Your Own Typhoon Cadet	106
Appendix D	Physical Quantities	109
Index		110

Chapter 1
Take Spare Boxer Shorts to Watch Robot Wars
By our war correspondent

Introduction

On a disused Air base about 10 miles East of Nottingham a man walks across the tarmac with his young nephew heading for a hangar with its doors barely open. Inside the hangar a war is being fought under the glare of the TV cameras. Robots are fighting each other to the death. Half way across the tarmac he is met by an RAF Officer about to take his Air Cadet Squadron to war. The Officer is as nervous as a long tailed cat in a room full of rocking chairs and informs our supporter that his Squadron robot 'Typhoon 2' is about to engage another enemy robot in combat.

Our supporter takes his seat in the audience only 10 feet from the side of the arena. Between him and the action are sheets of 12mm thick bullet proof Macralon. After watching a warm-up fight it is time for the semi-final he has come to watch. First in come the House Robots – Ref Bot, Cassias Chrome and Matilda. Then Typhoon 2 glides into the arena followed by its opponent X-Terminator. Typhoon 2 is shaped like a cone and painted as an RAF roundel. Its weapon claws look menacing and the shell and claws are slowly rotating.

An air of expectancy hangs over the arena as the production crew read out a list of conquests for X-Terminator which takes 30 seconds. In contrast, Typhoon 2 is an unseeded newcomer, and very much the underdog. The secret is out that Typhoon 2 is vulnerable while not spinning. Then….

3—2—1--Activate

There is a roar from Typhoon 2's petrol engine as X-Terminator makes a beeline for it in an attempt to cause major damage before the cone can spin up. Typhoon makes a quick defensive move behind the Ref Bot and X-Terminator sails past harmlessly. Now Typhoon is spinning well and a noticeable hum is hanging over the arena. The fight can really begin and for the next two minutes the robots are engaged in combat of the fiercest type, and the bangs (followed by Ouchs from the audience) come thick and fast.

Typhoon gives the vertical spinning disc of X-Terminator a huge swipe and knocks out its bearings. Weaponless, it hits the pit release button and attempts to lure its adversary down into the pit. Typhoon 2 uses this respite to gain energy and a menacing and intimidating whine fills the arena as the weapon cone spins faster and faster gaining more and more energy to unleash on its opponent. X-Terminator moves near to the arena wall in front of our supporter. Typhoon throws itself into X-Terminator and then hits the wall. A huge bang, clouds of dust and our supporter gets a very close-up view of a whirling Typhoon that he will never forget.

As he checks his Boxers a hush settles over the audience and everyone can hear the production crew shouting **"CEASE, CEASE, CEASE"**. The audience looks at the arena and as reality sets in, a deep breath is taken. Typhoon 2 has not only taken out X-Terminator, but also the arena wall just 10 feet away from our supporter. Typhoon 2 has destroyed 12 feet of bullet-proof Macralon and has also bent a steel arena support girder from vertical to 60 degrees, lowered the floor by 6 inches and 'O yes' destroyed a TV camera.

The filming was stopped, the arena made safe and the audience told to leave.

So what happened then?
Well, our supporter reports that Typhoon 2 returned to the war zone the following day and walked off with the Robot Wars **'Grand Champion'** Trophy and, by the way, took out another 6 feet of arena wall in the process.

JMW

****

Beginnings

It all started three years earlier in April 2001 when a Squadron of Edinburgh Air Cadets were discussing what to do as a project for the next year. On the white board their project officer, Flying Officer Peter Bennett, an ex RAF Test Pilot, had written three choices:

1. Build a wooden Unidentified Flying Object (UFO) for an exercise in the Pentland Hills near Edinburgh. The Cadets would guard it in gas masks and maybe someone would inform the local radio station..........

2. Build a radio controlled submarine with a Nessie monster superstructure for a covert test in Loch Ness at the height of the tourist season.

3. Build a fighting robot to compete in the BBC2 TV programme 'Robot Wars'.

After considerable discussion and much hilarity, the cadets chose Robot Wars by a narrow margin.

So began what was to become the most exciting and successful project ever undertaken by the Air Cadet organisation.

The First Brain-Storming Session.

Attack is the best form of defence so we need a good weapon.

The cadets analyzed the benefits of each type of robot design and each type of weapon:

Flippers like Chaos 2 were considered very successful, but they could do little real damage unless they tossed an opponent out of the arena. CO_2 gas bottles and pneumatics were also technologies that were very difficult and dangerous for inexperienced cadets to work with.

Axes like Dominator did not seem to do a lot of damage in the fights we had seen and pneumatic power was again a problem for us.

Battering Rams like Tornado were good, but considered boring. They didn't excite the cadets, as they seldom did a lot of visible damage and generally relied on brute force to push opponents down the pit.

Wedges are slightly better, but again do not do a lot of damage. They try to get an opponent's wheels off the ground and push him into arena hazards or the pit.

Pincers like Razer were attractive , but Razer is probably the ultimate of this type of design and we could not better it from our resources.

Small Spinning Discs like Pussy Cat looked promising, but again seemed disappointing in the degree of damage they inflicted.

Spinning Horizontal Drums seemed only slightly more dangerous than a small disc

Big Spinning Discs like Hypno-Disc inflicted real damage and were exciting to the cadets.

Many other weapons were suggested including projectiles, explosives, stun guns and flame throwers, but these were vetoed as they were against the rules.

What we wanted was a simple design which the cadets could build themselves and which was not too dangerous to construct in our hut with unskilled labour.

The cadets were really impressed with the damage that Hypno-Disc could achieve and we tried to think of a design that could do even better. Our military training teaches us that attack is the best form of defence and we wanted to build the most destructive weapon ever seen in Robot Wars.

Two of our Squadron Staff work for BAE Systems Avionics and suggested we try to apply some science and mathematics to the art of robot destruction. Damage required a release of energy so we should examine the equation of energy.

Rotary Energy = $\frac{1}{2}I\omega^2$

To get maximum rotary energy we therefore needed a high inertia ($\mathbf{I}$) (this requires a high mass concentrated at the edge with a large radius) together with a high speed of rotation ($\boldsymbol{\omega}$).

Should the weapon spin vertically or horizontally? Being an ex-Harrier pilot I knew a lot about the gyroscopic effects of rotating masses such as compressor fans. I therefore highlighted the problem that a vertical disc would act as a gyroscope, tend to prevent the robot from turning quickly and it could even topple over. In contrast a horizontal disc would allow the robot to turn quickly in the plane of the disc and resist any tendency to be toppled.

The idea surfaced for a heavy outer ring with cutters and this became the basis of our design. With the mass and radius fixed the Energy depended on the speed of rotation. This could be built up over several seconds like a flywheel and the stored energy unleashed on our opponent in a millisecond. In contrast, an axe could only build up energy in the half second the weapon took to accelerate from zero to its hitting speed.

How heavy should the weapon be? Half the weight of the robot was the general consensus. This rule of thumb has remained our aim throughout the project and has been applied to the complete Typhoon family with great success.

How big should it be? At least a metre in diameter and the bigger the better was the reply.

How fast should it spin? Several thousand rpm was the verdict, but it must spin up in less than 10 seconds and three seconds should be our aim.

We finished this first design session with the aim of building the robot around the weapon. The problems we were to meet later had not yet been thought of. We therefore went away in great spirits to put some detailed thought into our basic design concept.

The Second Brain-Storming Session.

Our second meeting had an agenda and looked like a meeting of the local council.

We had joined the Robot Wars Club and received a copy of the rules.

Three major problems had surfaced and had to be overcome to allow our design to work:

a.   How do we know which is the front of the robot when the cone is spinning.
b.   How do we incorporate a removable electrical link required by the rules.
c.   How do we self right if we get flipped over.

The brain storming came up with the idea of a flag to show the direction of travel and this could also be the removable link - Two problems solved in one.

Self righting was a major problem and many designs were examined to allow the robot to work both ways up. The problem was how to spin the outer shell without the structure holding the weapon ring interfering with the wheels. We also wanted a strong structure to hold the ring against major hits and this led us towards a central shaft. Next came the idea of a rotating cone to hold the ring, and thus the weapon could be the top armour as well. We had designed a full body spinner.

We still had no practical solution to right the robot if it got flipped over. We wondered if we could rely on the gyroscopic properties of a spinning cone to keep us level and decided to make a model to test this theory.

We needed a name and theme for our robot and suggestions came thick and fast. It had to be 'cool'. The girls wanted it covered in pink fur and call it 'Fluffy', but this was vetoed by the boys who said they wouldn't be seen dead with a furry pink fighting machine.

Being Air Cadets we turned to aircraft names and the Harrier and Eurofighter Typhoon were the most popular. As I was involved with the radar of the Typhoon and this name suggested a violent whirlwind, we agreed to call our robot 'Typhoon'.

We had great confidence that all we needed was to bolt some motors and metal together and turn up at the studio. Reality then dawned and we realised that we hadn't the skill or facilities to construct a serious robot and needed help. I approached my colleagues in the Radar Division of BAE Systems in Edinburgh and with the help of a 'Zero Pay – Own Time' job advert found several engineers and managers who were fans of robot wars and would jump at the chance of helping the cadets in their spare time. So began 'The Typhoon Robot Project' which was later to become 'The Typhoon Robots Project' as our family of robots expanded.

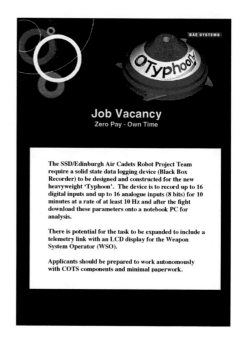

Job Vacancy
Zero Pay - Own Time

The SSD/Edinburgh Air Cadets Robot Project Team require a solid state data logging device (Black Box Recorder) to be designed and constructed for the new heavyweight 'Typhoon'. The device is to record up to 16 digital inputs and up to 16 analogue inputs (8 bits) for 10 minutes at a rate of at least 10 Hz and after the fight download these parameters onto a notebook PC for analysis.

There is potential for the task to be expanded to include a telemetry link with an LCD display for the Weapon System Operator (WSO).

Applicants should be prepared to work autonomously with COTS components and minimal paperwork.

Engineering

The right way to do engineering is to have a real goal and a tight timescale. This forces you to produce an efficient plan to achieve that goal. You design the hardware elements, make or buy the components, assemble the bits and test your creation thoroughly. The aim is to achieve the goal, on schedule and within your financial budget.

Disciplines and Skills

To build a fighting robot you must know a little about many different disciplines: batteries, motors, electronics, wiring, relays, computers, radio transmitters and receivers, control laws, belts, chains, sprockets, bearings, gear ratios, strength of materials, metals, plastics, machining, tapping threads, cutting, bending and welding etc. Team effort is needed with members knowledgeable in different aspects.

The instructors all had specific roles:

I was Team Manager and jack of all trades. I produced endless charts to plot our construction process and identify the critical path. These progress charts always started by showing a period of 4 weeks testing between the completion of the robot and the start of the filming. Unfortunately my calendar had more days allocated than the team had available and we always ended up with late night sessions in the final week to complete the robot on time. I also had a lathe and milling machine in my garage and produced many detailed parts for the project. I organised the cadets, looked after the finances and with the cadets, handled the publicity side of the project.

Martin Armistead is a radar mechanical engineer and was responsible for fitting all the parts of Typhoon into the space available under the cone. He is an expert in using CAD (Computer Aided Design). He also has a lathe and produced many difficult parts from some exotic materials. Martin is very determined and kept a cool head during our many crises.

Neil Harrison normally designs the digital parts of electronically scanned radars and relished the challenge of designing and making digital interfaces and high power motor controllers. He was always complaining that we designed the chassis first and left him no room to fit in his electronics. He would then turn up the following week with a perfectly fitting solution.

George Fisher, a Radar Chief Engineer was determined to cause maximum destruction. He was naturally promoted to the post of Typhoon Weapons Specialist and has masterminded the construction of all the weapon cones, rings and claws for the numerous Typhoons.

Tony Kinghorn is normally a Chief Radar Systems Engineer. He is a mathematician and boffin with a brain the size of a planet. He can do all the calculations in this book in his head and always comes up with a practical answer to several decimal places.

Roger Hill is a Chief Technologist and a dab hand at making printed circuit boards. He designed the special data logging 'Black Box Recorder' that has enabled us to verify the robot parameters described in this book.

The cadets who wanted to be seriously involved put their names on a list against a number of disciplines:

Project Manager	Weight Analysis
Assistant Project Manager	Manufacturing Plan
Mechanical Construction Team	Parts List & Procurement Plan
Electrical Construction Team	Financial Plan
Artistic Design Team	Test Plan
Publicity and Sponsorship Team	Training / Competition Plan
Safety Analysis	Data Logging and Analysis

Not everyone had to be a mechanical wizard and the genders were split 60% boys and 40% girls. The cadet project manager became Sergeant Graeme Horne (15) who is hoping to become an RAF Engineering Officer. His assistant was Cpl Alistair MacLeod who also drove the Lightning half of the Typhoon Twins clusterbot.

870 Squadron Cadet Robot Project Team
With Their Radar Engineer Instructors
Tony Kinghorn, George Fisher, Peter Bennett, Martin Armistead, Neil Harrison

The practical use of mathematics was needed to establish design parameters and this was the first time that some cadets had used mathematics for a practical purpose. A computer spreadsheet was used to determine the optimum weapon power and weight to achieve the maximum destructive energy in a realistic timescale. The use of high technology and simple designs were encouraged, such as the lightweight aluminium honeycomb base, the central safety break, concentric aerial and associated directional flag. These innovations were unique and overcame the problems that had prevented other roboteers from bringing similar design concepts to a practical conclusion.

The maximum weight limit required careful selection of materials and components involving a compromise in weight, rigidity and cost. Electrical engineering covered the full spectrum from radio control through digital low power logic circuits to high power motor controllers. Battery capacity and wire size were also important considerations.

Safety and reliability were built into the design from the start and test techniques established that included a number of current, voltage, speed and temperature sensors feeding an on-board data logger.

The cadets' artistic design skills resulted in the distinctive colour scheme and Typhoon logo which was modelled on that of the Eurofighter.

Model Tests and Strategy

Our first test model was made of Lego and a plastic bowl. It had Blutac cutters on the edge and showed that the concept of a cone weapon worked. The next model was a wooden cone with a low centre of gravity. This was easily flipped over and landed on its side more than 50% of the time. We then used an electric motor to spin up this model like a top and when we tried to flip it over it rose into the air and landed horizontally: **Eureka!**

Test Model

We realised that we would still be vulnerable before the cone spun up to a reasonable speed, but no one could come up with a practical self-righting idea. To prevent a flipper getting under our weapon ring we decided to incorporate ground skimming cutters to knock the flipper arm out of the way.

We had three months before the 5th wars and after a lot of heart searching decided we did not have enough experience to design and construct a heavyweight robot that could be competitive. We therefore decided to build a Middleweight (50kg) first in the hope that the competition would be less sophisticated. This turned out to be the best decision we ever made.

Construction Begins

The following week the cadets visited the local scrap yard and looked for a heavy metal wheel which could form the basis of our Typhoon weapon. After an hour of searching we came upon a large section of gas pipe about three quarters of a metre in diameter and several metres long. The helpful owner cut us off a small length with an acetylene torch and we found this weighed about 20kg. This determined the diameter of the robot and everything else stemmed from this dimension and mass.

The constant problem of weight was the most challenging process. We had to make many compromises in strength to get the total weight under the limit. Some parts, such as the central bearing housing, had to be machined and it was fortunate that Martin and I both had lathes in our garages.

We were given a large sheet of aluminium honeycomb material which is immensely strong and light. We used this for all our major chassis parts as it was easy to work with at our cadet headquarters without sophisticated tools.

We decided to fit two wheels to minimise weight and two powerful 1 horsepower wheelchair motors and two wheelchair control boards to allow us to dodge the opposition. A third similar motor was used to drive the weapon cone. To stabilise the base we used chair casters, but on its first test outing the casters broke so we replaced them with small fixed wheels.

We initially bought some big and reasonably cheap sealed lead acid batteries and had to change them at a late stage in Typhoon for smaller Hawker Cyclones to save weight. We had to do a similar exercise for Typhoon 2, but this time we went for Nickel Metal Hydride technology as it gave us the best power to weight ratio. Apart from the lathes we only used tools that were available in our cadet hut. Our local metal warehouse was a constant source of aluminum offcuts.

We bought a standard Futaba 40 MHz radio control from our local model shop and connected the servos to three potentiometers which connected to the motor speed controllers. This was quite crude, but very effective. For our later robots Neil designed an electronic interface that did not need mechanical servos.

Unlike the Eurofighter aircraft, we were not allowed any countermeasures which would disrupt the enemy radio control, but we managed to obtain some theatrical smoke cartridges and decided that it would be a useful countermeasure to lay down a smoke screen and hide if we got into trouble. We fitted 4 smoke cartridges round the edge of the base and controlled them from a separate Weapon Control Box which plugged into the main transmitter. If the worst happened and Typhoon was disabled, the Weapon Systems Operator would have let them all off simultaneously by pressing a 'self destruct' button to prevent the technology getting into enemy hands.

Smoke Cartridges

Middleweight 'Typhoon'

Being radar engineers there was a temptation for the instructors to think of the most complicated and exotic solution to a problem. Working with the cadets meant that everything had to be capable of being explained in simple scientific language and we therefore tried hard to implement the KISS (Keep It Simple, Stupid) design principle.

We also decided that where possible we would design our Typhoons with components that had an appropriate manufacturers specification and this stopped us implementing high risk design features such as overrunning our 24 volt motors with 36 volts.

Team Selection

Success in Robot Wars and Technogames requires not only a well designed and well constructed robot, but also considerable driving skill. An in-house competition was held amongst the 80 cadets in the Squadron to choose the best robot drivers and Weapon Systems Operators (WSO). Six cadets were shortlisted and a young new recruit Gary Cairns (13) won and became our primary driver. Gary's ambition is to join the RAF and pilot the real Typhoon. Practice is the key to success, and the cadets spent a lot of time tuning the control laws to achieve the best results.

We knew that roboteers were interviewed before each match so we gave our potential drivers and Weapon System Operators some practice interviews in front of a video camera. These were invaluable, and helped to avoid monosyllabic answers when being interviewed by Phillipa Forester. They also helped me to decide on the eventual team. The WSO for our first wars was Cadet Hazel Taylor (16) who raced go-karts in her spare time.

Cadets Hazel Taylor and Gary Cairns With the Middleweight Typhoon

Chapter 2
First Combat

The Pits at Elstree Studios

The day for recording dawned and we set out very early by car for Elstree studios in North London. The atmosphere in the pits was electric and it was an eye opener to see famous robots such as Chaos 2 and Hypno-Disc at first hand. As new boys we were very unsure of the procedures and no one told us we had to hand our radio transmitter in to the officials. We therefore received our first rollicking from Derek Foxley the technical chief when he saw it sitting on our bench. Over the next two years we were to learn that these frequent rollickings were his standard method of keeping enthusiastic roboteers in line. The next hurdle was the dreaded technical inspection. Would our novel combined flag and removable link pass scrutiny? Would the experts find some feature that transgressed one of their many rules and send us home to Edinburgh with our tail between our legs? We need not have worried, the technical crew were extremely helpful and actually complementary about our design.

Time dragged by and we learnt that there were two clocks used to control the filming process. The first belonged to the production company Mentorn and went very fast. If a robot was needed it was given a minute's notice to get ready. The second clock regulated everyone else in the pits and all there was to do was watch the hands crawl slowly round and round the dial as we waited for the call to action.

We utilised the long wait to visit our rivals and assess their robots for vulnerability and danger. Often we could identify a weak point in their armour which could be exploited in the arena. In chatting to these like-minded roboteers we got to know them quite well and established a rapport and friendship that is renewed each year. There is a great comradeship amongst the roboteers and they are always very happy to help out a fellow competitor who needs a particular component or special tool. Very occasionally we meet a team who is determined to win at all costs. They protest the judge's decisions, they protest the actions of the house robots and they protest the decisions of the production team.

We nervously waited for the pre-match interview, but when it came it was short and easy. What are your tactics? Who do you think is the weakest robot? Who are you going for first? How fast do you spin?

Entry into the arena is a complicated procedure designed to ensure safety in the event of a runaway robot or weapon. You push your robot into a pen. A low door is closed and you then lean over it to remove safety pins and insert the removable electrical link to activate your machine. An opposite door to the arena is then raised and you drive your robot into the arena and stop on your spotlighted position. Then you quickly climb the stairs to the control booth and await the start.

Awaiting the Start

Our First Fight

Our first middleweight fight was one of the most exciting in the history of robot wars with metal flying all over the arena. Typhoon rapidly spun up to its lethal speed. It fired two of its countermeasures cartridges and with white smoke billowing from under the cone it set about annihilating its opponents one by one. First to be destroyed was Zap whose armour was peeled off as it was knocked across the arena. Next was Doom which was opened up like a tin can. Typhoon then visited Mammoth and spectacularly amputated its legs. Across the arena to tackle Genesis, but they were no match for our whirling cone of destruction. Next to tackle the titanium armour of Hard Cheese. A major clash sent one of our cutters across the arena, but Hard Cheese was opened up to be eaten, which we duly did by bending his weapon disc and immobilising him.

Three of our five minutes had gone and five robots had been destroyed or immobilised by Typhoon and counted out by the Refbot. The House robots then moved in to dispose of them in the traditional way using the arena flipper, flame jets and the pit. Typhoon decided to engage reheat and tackle the house robots. Matilda was the first and in two mega clashes we knocked off her tusk and badly gashed her side. Poor Matilda tried to retaliate, but could not get her new disc weapon near the cone of Typhoon. She then attempted to push Typhoon into the pit, but we dodged out of the way and she almost fell in herself.

Robot pilot Gary then insists that he bumped into the Refbot by accident. Anyway, there was a huge clang, an even bigger cheer from the audience, and Refbot was knocked out and had to be replaced for the next round.

Hazel Proudly Holds Matilda's Tusk

At the post fight interview Craig Charles acknowledged we had a really serious robot and we issued a challenge to fight the military heavyweight robots which were due to compete in two days time. Unfortunately there was not room in the TV schedules for this to be arranged (or were they chicken?). Typhoon was 'Combat Proven' and the result was a climax to three months of very hard work.

We had managed to win the middleweight melee at our first attempt. The full body spinner concept worked and we had learnt a huge amount by being present at the recordings and seeing other robots at first hand.

Sponsorship At Last

The win gave us a huge boost and the Air Cadets great publicity. It also gave us a proven track record which we could use to get sponsorship and this helped us considerably as our cadet funds are not very large. Building robots is not cheap and our first Typhoon cost over £1000. Financial support was essential, but we found it easier to ask firms to provide us with material parts such as high power connectors or specific services such as welding. We now have over 20 sponsors (Appendix A) and are most grateful to them for their support and encouragement for our project.

Press Day

After returning to Edinburgh our Air Cadet publicity machine went into top gear and it was arranged for 'Typhoon' to be flown into our Edinburgh squadron headquarters by helicopter to a hero's welcome in front of the press. It took pride of place on the parade square, led a marchpast and was enrolled as an 870 Squadron ATC Cadet. At only two months old it was by far the youngest cadet in the Air Training Corps! It has already been awarded an RAF Marksman badge which it wears very proudly.

Typhoon Leads The March Past

Over four million people have seen our 'Typhoon' robot win the middleweight championship on television and there has been considerable press coverage. The Chief of the UK Air Staff also sent a personal congratulatory letter to the Squadron.

An RAF Marksman's Badge For Typhoon

When you have a winning design don't change it. We therefore scaled up our design the following year to a heavyweight size (Typhoon 2) and scaled it down to build two lightweight versions (Typhoon Thunder and Typhoon Lightning).

Chapter 3
Technogames

At the end of 2002 our 'Typhoon' team entered the BBC programme 'Technogames'. This is a spectacular televised robot Olympics for which teams design and build technological devices to compete in a number of sporting events. The cadets competed in the assault course (where two robots race round obstructions, over ramps and charge through a tyre wall), the football competition (where two teams of two robots compete against each other to score the most goals) and the Cycling event. The 10 programmes were broadcast on BBC2 just before Easter 2003.

Assault Course

Two machines compete head to head in a timed trial over a course of approximately 25 metres, testing the speed, control and manoeuvrability of each machine. The machines race through a series of obstacles including a barrel, seesaw and a tyre wall. Time penalties are incurred if a machine knocks over or misses out any obstacles.

'Typhoon Rover' was specially designed for the assault course with its wooden chassis optimised to go up ramps and steer a football into a goal. Typhoon Rover contains the main parts of our heavyweight fighting robot 'Typhoon 2' and is powered by four 1 Horse Power electric motors with independent drive to each of its four wheels.

Typhoon Rover
Pilot: Cadet Gary Cairns
Tactician: Cadet Jonna Salvesen

'Typhoon Rover', piloted by Cadet Gary Cairns, was fantastic. It gently steered round obstacles and scored a penalty goal first time in every heat. After a hard fought and nail-biting competition we won the Gold Medal with a new World Record time for this event.

Courtesy Mentorn

Rover Approaches The Technogames Ramp

Football

Two teams with two robots each compete in a match lasting 5 minutes. The winner is the team scoring the most goals. The ball is a ten pin bowling ball with football boot studs all over it to stop it rolling too fast across the arena. It looked like a porcupine.

'Team Typhoon' entered a pair of robots in this event: 'Typhoon Rover' and '870' (Our Squadron number). '870' is actually 'Typhoon' our Robot Wars middleweight champion fighting robot with its weapon cone inhibited and fixed. We decided to have a dedicated goalkeeper and therefore fixed large white hands to its front and back. The overall length of 1.6m was well within the 2 metres allowed. Because it had previously appeared in Robot Wars we had to call it a different name and give it a new colour scheme (silver with two RAF roundels).

Typhoon Rover (Centre Forward)	**870 (Goalkeeper)**
Pilot: Cadet Gary Cairns	Pilot: Corporal Andrew Fullerton
Tactician: Cadet Andrew Snedden	Tactician: Cadet Naomi McGeary

Captain: Sergeant Graeme Horne

23

'Team Typhoon' started extremely well, winning our first two matches 6-0 and 1-0. The semi final was a highly physical match which we lost 2-1 but gained **Bronze Medals** as a result. Next year we hope to construct a new robot goalkeeper for the football. '870' is on the transfer list as it was difficult to manoeuvre and easily pushed around due to its light weight.

Cycling

The cycle race is around a flat oval circuit, approximately 50 metres in length. Cycles are timed over a set number of laps and must be self balancing. Devices such as gyros are allowed, so we fitted a large instructional gyroscope to the frame of a small child's bicycle and made a closed loop stabilisation system using a pendulum and two large servos to move the handlebars. We called our bicycle 'Byphoon'.

Initial trials in our car park showed a bigger and bigger (divergent) oscillation of the pendulum. To improve the stability control loop we therefore shortened the pendulum to reduce the time constant and damped it using washing up fluid in an aluminium pot.

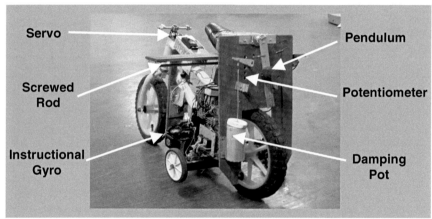

'Byphoon'

During practice the night before our race, one of the two large servos driving the handlebars failed. The remaining single servo was only able to turn the handlebars at a slow rate which changed the time constant of the stability servo loop. We had no spare large servo and had to compete with a half serviceable system. The result was not unexpected and our bicycle left the track halfway round the first curve and crashed. Nevertheless, the cadets learnt a lot about several aspects of engineering and physics from this bicycle stability project and want to have another attempt at the next Technogames.

Technogames Studio

As we had entries in three Technogames events, ten 870 Squadron cadets spent four days at RAF Newton for the filming. Cadet Jonna Salvesen (14) was the Assault course tactician and reports:

Cadet Report on Technogames

"We took three robots to the studios at Newton where Technogames takes place in a huge aircraft hanger. Our robot "Typhoon Rover" was to take part in the assault course, and after making it through to the semi-finals, our skilled driver Cadet Gary Cairns drove a perfect run of the course with absolutely no time penalties.

We had made it to the finals against 'Mighty Mouse' and the pressure was on. Rover knocked the heavy oil drum aside and slid through the gap. Gary deftly manoeuvred the football into the centre of the goal and we were first up the ramp. As Rover burst through the tyre wall and crossed the finish line, we had won. A tremendous applause came from the crowds, and we could see the Air Cadet banner being raised by our faithful supporters. The overall time for that run came out as 10.5 seconds which, we later found out, was a new world record!

Our Commandant, Air Commodore Chitty, had come to support us that day. We were honoured to have him present us with gold medals at the medal ceremony.

Overnight there were a few adjustments to be made to Rover to prepare it for the football competition. We also fixed two big hands onto our conical goalkeeping robot '870'.

At the studio, we won our first match against 'Team Pink' with a score of six to nil. Our quarter final opponents were the 'Toasters' and after a heart stopping head-on clash at the centre spot which sent the ball flying dangerously high, we manoeuvred the ball into the net and won by that single goal.

In the semi-final we were up against our roboteer rivals "Big Bro". The whistle blew and we began. It was a tense match, and the supporters stood at the side lines biting their fingernails as 'Rover' tried hard to find a gap. With the score at one goal each, all four robots were locked in a pushing match at the side of our goal. It was a case of who gave way first and sadly Big Bro got the ball past '870' in the last few seconds of the game. But we did put up a good fight.

The competition had been tough and very close, but we were very proud that we had made the final three. We walked away that day with three gold medals, six bronze medals and a new world record. So all of our hard work and preparation had paid off.

All of the cadets feel that this has been a wonderful and rewarding experience, We didn't expect the robot project to come this far and do so well. We are all proud of the teams, the supporters and our instructors. This has been a great achievement for all of the staff and cadets of 870 (Dreghorn) Squadron.

Cadet Jonna Salvesen

Chapter 4
Off To War Again

The Family Expands

Our teenage air cadets are immensely proud of having helped to design, build and operate such successful machines. The 'Typhoon' family started with the middleweight (50kg) prototype. This was followed the following year by its big brother; a heavyweight (100kg) 'Typhoon 2'. The cadets also produced two lightweight (25kg) look-alike twin robots ('Typhoon Thunder' and 'Typhoon Lightning') for both competition and training. Thunder was entered in the Lightweight championship on its own, but both lightweights were teamed up as a clusterbot as an additional entry in the Middleweight championship. We believe this was only the second clusterbot in the history of Robot Wars. Driving the 'Typhoon Twins' into the arena as a single unit required similar skill to a close formation flypast in a pair of Eurofighter Typhoons.

The influence of the female cadets can be seen in 'Typhoon Thunder' the robot they built themselves which is fitted with a pleated mini skirt.

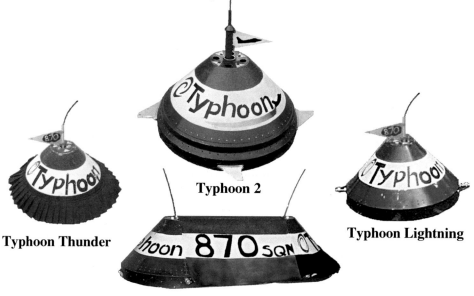

Typhoon 2

Typhoon Thunder

Typhoon Lightning

Typhoon Twins

Robot Wars Extreme 2

All four Typhoons were entered in the 2002 Robot Wars Extreme 2. As we had entries in three events, nine 870 Squadron cadets spent five days at RAF Newton for the filming. Craig Charles was again the main presenter and Phillipa Forester conducted the pre fight interviews. The cadets had passes to the pits and were able to see famous robots such as 'Chaos 2', 'Tornado' and World Champion 'Razer' at close quarters. They had a special photo session in the arena with the new house robots and were invited into the house robot control box for some of the fights. They even drove a house robot themselves between events.

Typhoon Twins

The cadets were keen to join Thunder and Lightning together and enter them as a clusterbot in the middleweight competition. The rules stated that they must enter the arena as one unit and only split up after the start. We thought of many ways of joining the two cones together and eventually decided on the simple solution of tying some wire between the claws. The two cones did not look like one robot so we covered them with paper and this came off easily when the two separated and the cones spun up at the start of the fight. Our initial attempts to drive the clusterbot in a straight line and do a gentle turn ended in disaster and broken wires. After some experimenting we hit on the solution. The lead robot 'Thunder' (Ladies first into combat) did the steering and determined our path into the arena.. Lightning provided the motive power for both sections and pushed Thunder ahead of it so that the wires were not under tension and did not break. The driver of Lightning also had to keep his flag pointing straight at Thunder.

In our first heat Lightning was unfortunately over the pit when it went down. Fortunately the rules had been changed that year so that more than 50% of the clusterbot had to be immobilised to eliminate it. The rule was changed back the following year to 50% or more immobilised. This was a pity and does little to encourage clusterbots which are liked by the audience.

The two lightweight robots are very useful for demonstrations as they are not very heavy or powerful and therefore reasonably safe provided the cone motors are disconnected. We built a simple maze about 4m by 3m with wooden kerbs and this has gone to fetes and also to the Edinburgh Science Festival. The public love driving these small robots around the maze and some of the small children are very good at controlling them.

**The Public Drive 'Typhoon Lightning'
At The Edinburgh Science Festival**

Extreme 2 - Fight Reports

LIGHTWEIGHT CONTEST

"The lightweight fight was straight forward and our opponent was easy meat for 'Typhoon Thunder' which was made by the girl cadets. Its pleated mini-skirt wasn't even torn and Cpl Keri Scott showed the audience that agression was not only for the boys."

MIDDLEWEIGHT COMPETITION

"Our 'Typhoon Twins' clusterbot of 'Thunder & Lightning' was not expected to last long in the Middleweight competition due to their light weight. They sustained severe damage to their cones from a formidable robot with a vertical spinning disc called 259, but just managed to survive their heat. In the final the 'Typhoon Twins' were up against their big brother 'Typhoon' and faced the vertical spinner and another powerful and heavily armoured robot 'Steel Sandwich'. The tactical plan was for 'Typhoon to mount a frontal attack on the vertical spinner whilst the Twins executed a pincer movement on the other adversary. The plan worked perfectly; 'Typhoon' spun up to maximum speed and gave 259 a huge horizontal swipe; the target's vertical spinning disc acted as a gyroscope and the gyro torque reaction caused the robot to invert instantly. It took no further part in the fight. 'Typhoon' then went to the assistance of the 'Twins' and together they disabled the remaining opposition. The game plan was then to fight each other to the finish, but all three Typhoons had sustained battle damage and were unable to deliver a killing blow. The fight therefore went to a judges decision. On the basis of the four criteria: Style, Control Damage and Aggression , after much deliberation, the decision went to 'Typhoon'."

Sgt Graeme Horne WSO

Our heavyweight robot 'Typhoon 2' unfortunately exited the Annihilator contest after inflicting huge damage to 'Kan Opener' the eventual winner. It was unlucky to be flipped over when its cone was stationary after being caught on the claw of its adversary. That, however, is what makes Robot Wars such an interesting and unpredictable contest.

The 7th Wars

The 7th Wars were held in late Summer 2003. We entered the main heavyweight competition with Typhoon 2 and also the Middleweight and Featherweight events. There was no Lightweight competition that year.

Typhoon Cadet

Typhoon Cadet, a 12kg Featherweight, is the newest addition to our family. Was it a coincidence that our baby cadet first ran round our parade hall exactly 9 months after our clusterbot appeared in the robot wars arena? Typhoon Cadet uses 3 electric drill motors; two for the wheels and one for the weapon. It provided us with a huge challenge to get the weight below the limit and we had to drill lightening holes in every piece of metal.

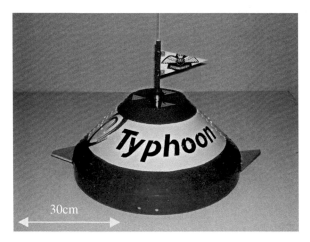

Typhoon Cadet

Typhoon Cadet has a sporty performance and is extremely popular with the cadets and the girls in particular. It is not quite so dangerous as the heavier robots and is easily lifted and transported. There was great competition amongst the cadets to be part of the featherweight team for the 7th wars.

Typhoon 2

The Typhoon 2 robot engineering and systems have some similarities to Eurofighter. The chassis is made from aluminium honeycomb; the outer cone armour is fabricated from titanium sheet. Separate electrical systems are incorporated to minimise single point failures following battle damage.

The heavy steel outer ring and cone is powered by a chain saw petrol engine and rotates at up to 1000 RPM with the weapon cutters travelling at over 110 mph. The cutters are made of Hardox, an extremely hard Swedish steel that is normally used for the teeth of digger buckets. The weapon cone acts as a huge gyroscope so no self-righting mechanism is fitted. Horizontal acceleration and manoeuvrability are therefore important design parameters as all the 'Typhoons' are vulnerable to being flipped over during the first few seconds of the fight before the cone spins up and reaches gyroscopic speed.

The most unusual component is the high-tech device designed to eliminate wobble of the cone at high speed. This is called the 'Wobulator' and its design is a close kept military secret.

Making Typhoon 2

As Typhoon 2 was unseeded we had to attend the qualifying event. This was almost a disaster as we failed to spin up the weapon cone to a fighting speed (the reason is too embarrassing to record) and when we attacked another robot the cone stopped and we were unceremoniously flipped over by Ewe 2 and counted out. We had a miserable weekend believing we had not qualified, but on Sunday afternoon we got the call that we had gained a discretionary place. This is generally code for being wanted as interesting 'Cannon fodder' for other robots to destroy. Nevertheless we went down to Nottingham again in high spirits with a minibus full of cadet supporters. Their vocal support in the audience was magnificent and led to them being nick-named 'The Scream Team'.

7th Wars - Fight Reports

FEATHERWEIGHT HEAT:

It was hugely disappointing that as soon as the fight started Typhoon Cadet began to make uncommanded movements and its weapon cone kept stopping and starting. We believe this was due to electromagnetic interference, but do not know where it came from, or if someone was transmitting on our frequency. Typhoon Cadet was unceremoniously pushed onto the arena flipper and its maiden flight was short, but spectacular. Fortunately it suffered only minor damage and will live to fight again.

MIDDLEWEIGHT CONTEST

In the Middleweight Melee our opponents were Steel Sandwich and Phoenix. As our cone began to spin up we saw smoke coming from the edge and feared the worst. Fortunately this was only a slipping drive belt and by limiting the use of the weapon motor we were able to overcome the problem and knock out Phoenix. We then hit Steel Sandwich and jammed one of its wheels. As it frantically tried to squirm away from us we stopped our cone and pushed him into the pit.

Cadet Tina Bowman - Weapon Systems Operator

HEAVYWEIGHT - HEAT O

Round 1 - Heat Melee

We were devastated to find out that a seeded robot, the famous flipper Bigger Brother, would be in our heat. Typhoon 2 had a technical problem with the starter free wheel which prevented us starting the petrol engine and spinning up the weapon. We initially managed to dodge our opponents, but Bigger Brother was on top form and after disposing of the other two robots - URO and Colossus, he eventually flipped us over. Fortunately we were the last to be immobilised and went through to the second round, which was a surprise for a "cannon fodder" robot.

Round 2 - Heat Semi Final

Hammer Head 2, another flipper, were our opponents in Round 2. We finally got our starter freewheel to work and unleashed Typhoon's power. We gracefully dodged our opponent while tactically gaining speed and energy in our Weapon of Mass Destruction. Then we struck. This was the first time that Typhoon 2 had hit an object while in "super cruise" and it was good! We sent Hammer Head 2 bouncing round the arena like a game of pinball and immobilised him. It was only after the fight that we saw how much damage we had caused to its rear armour. We had sliced like butter through two layers of 15 millimetre thick sheets of Polycarbonate and gashed the 6 millimetre aluminium under that. As the clear winner we were through to the heat final.

Heat Final

The tension was building in the pits, it was beginning to sink in that our robot was doing exactly what it was designed to do – damage other robots. This time we were up against Iron Awe 2.1 which was armed with both an axe and a flipper. Again we elegantly drove into the arena with the engine idling. With another textbook spin up we effectively engaged and knocked the enemy all over the arena. We returned to base with little more than a scratch, although the team decided Typhoon 2 could do with a touch up paint job.

SEMI-FINALS

Last 16

Things now started to get serious; we were now guaranteed to be seeded in the next wars however we performed in this battle. Thermidor 2, our opponents were intimidating. They must also have a good robot to get this far, but it was the same routine. A tactical retreat to get valuable spin up time before unleashing the large amount of energy in our spinning disc. On impact we did very little visible damage, but due to the nature of their robot we actually broke their batteries and drive therefore immobilising them with our first hit.

Quarter Final

Atomic was a powerful flipper and persistently drove into us and attempted to flip us over. After several knocks our disc slowed down and they were able to push us into the arena side, although they failed to flip us. We managed to escape from their grip and spin up properly. With the disc spinning nicely we got a good hit and penetrated their titanium armour. They were knocked out and were a sitting duck, at this point we went for a victory lap around the arena although a shout came through from Craig Charles requesting another hit, we agreed but felt it was not very sporting to hit them after they had been immobilised.

Semi Final

Our opponents for the semi final were X-Terminator. After some initial blows without too much damage we were given a chance to spin the disc all the way up to reheat. The disc was going faster than it had ever been before. Gary lined up and gently glided Typhoon 2 in for the knockout punch. After hitting X-Terminator we unfortunately drove into the side of the arena shattering two panels of bullet-proof Macralon and bending a steel support girder. Immediately the fight was ceased, and the production team ran about working out what to do. Nothing like this had ever happened before in Robot Wars. The robots were quarantined over night so as the fight could be continued in the morning. This gave us an opportunity to see we had knocked the bearings out of X-Terminators' vertical spinning disc, reduced the arena wall to small pieces and lowered the floor by 6 inches. We resumed the next day and continued our destruction knocking out the drive from one side of X- Terminator.

Flt Sgt Graeme Horne - Weapon Systems Operator

The Grand Final

This was the day we had dreamed about since Typhoon first appeared in the middleweight class three years ago. Typhoon 2 was in the Robot Wars UK Grand Final. Our opponents were the mighty Storm II, another very well designed robot with one aim, to destroy the enemy.

Storm II was very fast and agile which made it hard to build up momentum in our weapon disc as every clash slowed it down. Although we were not spinning very fast we hit and holed the arena wall again, it was not as spectacular as before, but the fight still had to be stopped while the production crew replaced the bullet-proof panel. After it was fixed we were unable to get the petrol engine started again, the tension was huge, if it was not started we would be a sitting duck. Looking rather weak, we trundled into the arena with a dead engine. Just before the 3-2-1-Activate we tried the starter again and the engine fired although it cut out again shortly afterwards. We were pushed into a corner and lifted up several times, but not flipped over. After a few more knocks from Storm II we finally got the engine started and managed to rip a titanium armour panel off its front along with some internal structure. The fight went to a Judges Decision, Storm II won points for Control and Aggression but Typhoon 2 won points for Style and Damage. Due to the weighting behind the points, it meant Typhoon 2 won the contest and became **ROBOT WARS UK SERIES SEVEN GRAND CHAMPION.** Cpl Gary Cairns - Driver

Cpl Gary Cairns, Sgt Graeme Horne and Fg Off Peter Bennett with Jayne Middlemiss, Craig Charles and the Grand Champion Trophy

If you watch Typhoon 2 fighting you may see some rings of light near the top of the cone. These are high brightness diodes which show the Weapon Systems Operator how fast the cone is rotating. His weapon control mimics the Eurofighter Typhoon throttle positions of 'Cruise – Supercruise – Reheat' and the coded lights are a valuable indication of our energy and damage potential. We may increase the number of lights for future wars to give us more precise information.

As an ex RAF test pilot I am a great believer in analysing our robot's performance in every fight. We therefore built a black box fight recorder which records 16 Parameters including speeds, current, voltage, temperatures and shock. It has been an invaluable tool to help us optimise our design.

Our starter motor freewheel gave us major problems throughout the 7th wars as it kept freewheeling in both directions. It was very frustrating to have a working engine that we couldn't always start and we only just managed to start it in the final.

We are extremely pleased with Typhoon 2 and never imagined it could do so well. Self damage was our biggest concern as every action has an equal and opposite reaction. We finished building it too late to do more than a simple spin-up test, so the arena was its testing ground. We kept the rotation speed down for our early fights, but when we found we were undamaged we became more confident. We increased the weapon power from cruise to supercruise and then in the semi final selected reheat. The fights became more and more spectacular and the sparks coming from the titanium armour of our opponents looked great. It was exhilarating to feel the power under our control and hear the intimidating whine of the petrol engine as it spun up the weapon to lethal speed.

It was an unbelievable feeling to get through to the finals. We were excited, surprised and over the moon, for we never once thought seriously that we would get this far. We had reached the final against all the odds and put the Air Cadets on the map. Before the fight Sergeant Graeme Horne and Corporal Gary Cairns were as cool as cucumbers. The worst moment was the unusually long wait in the final between the "Roboteers standby" and the countdown. During this time Gary's heart was pounding harder than Mr Psycho's hammer as the realisation finally dawned that we were in the final. However, once the "activate" call came his cadet training took over and he was fine.

The final is a huge blur in our memories, but Storm II was a worthy opponent and it was extremely difficult to avoid him long enough for our cone to spin up properly. Once the battle started Gary just kept cool and treated the fight as a computer game.

It was overwhelming to experience the awesome power and energy of the machine we had created. The audience were screaming for Typhoon and the whole arena including the control box moved when the robots clashed. The cheers of our team of 870 Squadron supporters in the audience could be heard clearly in the control booth.

This was a huge source of encouragement to the team as it was like playing a cup match at your home ground.

870 Squadron Supporters With Typhoon 2 in the Pits

The emotional experience of winning was incredible and for several days afterwards we had to pinch ourselves to confirm it was not a dream. We were very lucky to survive so long, because we had brought very few spares with us. We had two sets of batteries and after the final, one set was too hot to recharge immediately and the spare set was in our medium weight Typhoon which was waiting to fight in the middleweight final. We borrowed some ice from the producers fridge to cool the batteries down. We eventually arrived home elated at 2 o'clock on Monday morning and dragged parents from their beds in dressing gowns to see the Grand Champion trophy.

We are keen to defend our title in the next wars, but it will be a difficult task as Typhoon 2 is now the Champion to beat. The country's best roboteering brains will be devising Typhoon defeating tactics. Nevertheless we will try to stay one step ahead and have started on a Typhoon 2 improvement programme which will address the few weaknesses we identified during the 7th Wars.

Our first middleweight robot was completed in 3 months. Typhoon 2 took 12 months, but we were also building the two lightweights at the same time. Only once during construction did we consider giving up - When Typhoon was well over weight and there was no obvious way of reducing it without compromising the mass of our weapon. It was team spirit and the desire not to let the cadet project fail that kept us going.

Honours

Was it worth it? Our cadet squadron trophy cabinet is a testament to our success and it now has five 'Robot Wars' championship trophies and Technogames Gold and Silver medals. The last one had to be purchased ourselves as due to an 'oversight' by the production company Mentorn, they failed to provide one for the Middleweight Championship.

The cadets have really enjoyed building, testing, driving and supporting their robots and they and their instructors (who are BAE Systems radar engineers) deserve the credit for a fantastic job well done. It has been a real team effort and the cadets have learnt that engineering can be real fun.

Honours gained so far
Robot Wars Middleweight Champions 2001
Robot Wars Middleweight Champions 2002
Robot Wars Lightweight Champions 2002
Technogames Assault Course Gold Medallists 2003
Technogames Football Bronze Medallists 2003
Robot Wars Middleweight Champions 2004

Robot Wars UK Grand Champions 2004

Chapter 5
Weaknesses Must Be Minimised

Basic Design

No robot design is invulnerable. There are always weaknesses, but these depend on the design features of your opponent. Robot combat is like the game: 'Paper - Scissors – Stone'. Scissors can cut Paper; Paper can wrap Stone; Stone can blunt Scissors. So it is with robot design. Some designs can defeat other designs, but it is very rare for one design to come out on top against every type of opponent. Roboteers will soon come up with a 'Typhoon'-defeating idea and incorporate it into their design. One idea already used against spinners is a large wooden block fixed to the back of a flipper. The flipper backs into the spinner and accepts damage to the wood as it absorbs the spinner's energy and stops its disc. Before it has time to spin up again and replenish its energy, the flipper does a quick about turn and uses its flipper to turn over the spinner.

Ideally we want to knock out our opponent with the minimum number of hits. We should target exposed parts of his weaponry and wheels which can be bent or broken off. Our most difficult opponent is therefore a solidly built ramming robot or wedge which has concentrated all its weight allowance in thick armour. This can take repeated big hits with little damage and there is a high likelihood that the spinner will break itself.

The outer ring of Typhoon 2 is also the main armour and takes the majority of the hits from enemy robots. The titanium outer cone is angled at 45 degrees which makes it quite difficult for vertical spinning discs to touch. Even axes have difficulty penetrating a curved surface angled at 45 degrees. A gripping robot like Razer or Sir Killalot will also have difficulty getting hold of a cone.

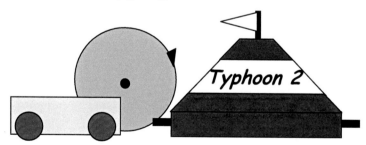

Newton's Third Law states; "To every action, there is an equal and opposite reaction." In other words if you hit your opponent with a force of X Newtons an equal shock of X Newtons will react on Typhoon. The chassis therefore has to be extremely strong to survive and all the components must be securely fastened down to avoid them breaking away. We were very concerned that the magnets inside our wheel motors might detach so we mounted the wheel boxes on rubber to absorb and reduce the peak shock reaching the motors.

Self-Righting

We have no self righting mechanism at present and our major vulnerability is being flipped over. If we are not spinning this is quite likely and has already happened in two competition fights. If we are spinning at a reasonable speed we rely on two features of the design. The first is the two ground skimming cutters which are designed to knock flippers aside. The second is the gyroscopic properties of the spinning weapon/cone.

How a Gyroscope Works

If you remove the front wheel of a bicycle, hold the ends of the axle in your hands and get someone to spin the tyre, you have a simple gyroscope. If you tilt the axle, the wheel will tilt in another plane. Gyroscopic theory states that if you apply a force to tilt a gyro the reaction will be experienced 90 degrees round the gyro in the direction of rotation of the wheel. Therefore if you apply a horizontal force to a vertical spinning wheel it will react in the vertical plane and turn over.

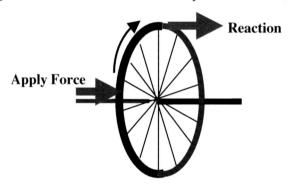

In Robot Wars Extreme 2 Typhoon gave the vertical spinner 259 a huge horizontal swipe. Due to the gyroscopic effect of its wheel 259 turned over onto its back and was counted out.

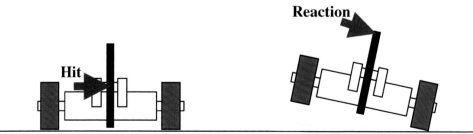

Vertical Spinner Receiving Horizontal Hit

We thought of constructing our superstructure as a hemisphere which would roll back upright if it was flipped. In actual fact the top would need to be more than a hemisphere as the centre of gravity of Typhoon is about 100mm off the ground and the equator of the sphere would need to be above this centre of gravity. The problems in making a strong sphere were considered beyond our capabilities.

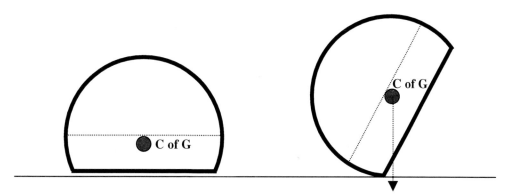

Centre of Gravity Below Equator of a Sphere Will Cause It To Roll Upright

We are currently working on a novel self-righting mechanism to be incorporated in Typhoon 2 for the next wars. At the time of writing the device has not been tested and is still on the secret list.

Chapter 6
Depleted Uranium, Titanium or Unobtainium

Construction Materials

If our robot's skin is ripped off or its chassis or internal bracing fails we are doomed. We therefore need a very strong structure which must also be very light. Different metals have different characteristics of strength, stiffness, hardness, ease of bending, and ease of machining. The chassis must be light and stiff. Internal webs must be light and easy to bend. At the other extreme the weapon claws need to be extremely hard and their mass is not so important as they contribute to the inertia of the weapon cone.

Aluminium

Aluminium is a very popular constructional material as it is light and reasonably strong, though it is definitely not as tough as steel. We used it extensively for our chassis and also for the cones of our middleweight and lightweight Typhoons. There are several different aluminium alloys and you can purchase plate in different thicknesses and obtain extrusions such as angle, tee shape, I-beams, C channel and round tubing. It is easy to work with, but requires specialist equipment to weld it.

When we started to design our first Typhoon we knew that weight would be critical. I remembered that a modelmaking friend had once extolled the virtues of aluminium honeycomb material to me so I contacted him and obtained some large off-cuts that proved perfect for our chassis. This material is used extensively in aircraft construction and has a huge strength to weight ratio. Another bonus was that it was easy for the cadets to use without sophisticated tools. It could be cut into complicated shapes using a standard jig saw, and by cutting away a short depth of honeycomb interior with a Stanley knife, the sides could be bent outwards to form flanges that could be bolted easily to other structural components. This material was easy to use as it did not require great strength to bend once the honeycomb had been cut.

Flt Sgt Graeme Horne
Cutting Honeycomb Webs

Steel

Steel is tough, cheap and can be welded easily. It can also be obtained in sheet and extruded sections, but it is heavy. We used it for the weapon ring which was also our armour, but only used it on the chassis when absolutely necessary.

For the middleweight Typhoon's weapon ring the cadets were fortunate in getting a small length of steel gas pipe from our local scrap yard. The rest of the robot was then designed around this ring dimension. When we came to build two lightweight versions we were not so lucky and therefore decided to manufacture our own ring. First we cut out three circles of MDF (Medium Density Fibreboard) material and screwed them together with suitable spacers as a central former. We then purchased 12 strips of 1mm thick steel plate which was 50 mm wide. We wrapped the first strip round the former, cut it to length (the exact circumference) and applied glue to the outside. We then wrapped the second strip round the first ensuring that the join was well away from the first. A nylon strap was ratcheted up tight round the outside to hold the strips together whilst the glue dried. The process was repeated each parade evening with another strip of steel and we eventually had a 12mm thick laminated steel ring of 600mm diameter. It looked like plywood made of steel. A row of 3mm steel bolts at intervals around the ring ensured it would not delaminate in action. The cutters were attached with 16mm bolts inserted through the laminated ring from the inside and stabilised by triangular shapes of 6mm thick steel plate.

Laminated Steel Weapon Ring

For Typhoon 2 we were offered a specially turned steel ring and this has saved us a huge amount of work and provided us with a weapon ring that did not require much balancing.

For the weapon claws we have used Hardox the hardest steel we know of. It is made in Sweden and is normally used for components such as digger buckets. We were put in touch with a local firm which specialises in this material and has the facilities to cut it underwater with a thermal lance to retain its hardness.

It is very important for safety reasons that the claws don't break off when they hit another robot. For Typhoon 2 we therefore had them welded to the main ring and got a colleague to check that the strain from a major hit would be distributed evenly over the weld and not concentrated in one area that could be overstressed and fail. This finite element analysis enabled us to adjust the shape to ensure that the claws would remain attached.

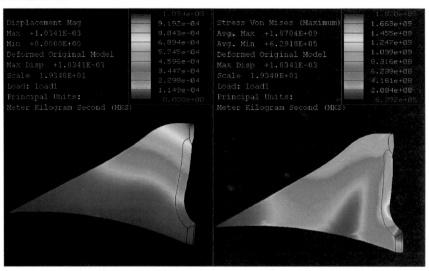

Displacement   **Stress**

Finite Element Analysis of the Claw Attachment

Titanium

Our original Middleweight Typhoon had an aluminium cone and this was repeated for the lightweights and featherweight versions. Aluminium was adequate, but was easily damaged and shows considerable scars from winning 3 championships. The various nasty spikes sticking out of our Mammoth adversary (when we amputated its legs) and the formidable vertical spinning disc of 259 have all left permanent scars. For Typhoon 2 we therefore decided to use titanium.

Titanium is probably the strongest metal around. It is heavier than aluminium and much more expensive. It is used extensively in military applications requiring lightweight armour and for containment rings for jet engines. It can withstand deformation and bending much better than aluminium or most steels, but it is very difficult to cut and drill. We thought it would be an ideal material for the cone of Typhoon 2, but we needed the seam to be welded so that opponents weapons would not catch on the join and rip it open. Titanium will catch fire at extremely high temperatures so it must be welded in a chamber filled with inert gas. This is expensive, but fortunately BAE Systems offered to weld the titanium cone for us at their Eurofighter Typhoon factory at Samlesbury.

In its first fight (The Annihilator) Typhoon 2 broke the central shaft of Kan Opener and rebounded into its claw which penetrated our cone. The titanium thickness has subsequently been increased to a new secret thickness to prevent this happening again.

Material	Ultimate Strength (N/mm^2)	Density (kg/m^2)
Aluminium	90	2700
Steel	380	7850
Titanium	552	4420

We wanted to use a vee belt to drive the cone of Typhoon as we needed something that would absorb shocks from the weapon without transferring them to the electric motor. For the wheels we used bicycle chain as this was easier to drive from a small sprocket to achieve a reasonable drive ratio. We calculated (Chapter 7) that they would survive the shock loads without breaking.

Shock Absorbers

To minimise the peak 'G' shock experienced on our own chassis during an attack we mounted our vital components, such as the wheel boxes, drive motors and batteries, on rubber shock absorbers.

Design Tools

Our middleweight, lightweight and featherweight robots were all designed using pencil and paper. For Typhoon 2, however we knew that it would be extremely difficult to fit all the components into the small conical space available and therefore Martin used 3D CAD (Computer Aided Design) tools to draw the basic shapes. This was extremely useful as he could change the thickness of the weapon ring by one millimetre and instantly see the effect on the overall weight. Some of the design was done with our cadets in the Squadron HQ with the computer CAD picture projected onto the wall.

Despite our confidence in these computer tools we still made a wooden template of the internal section of the cone so that we could check the clearance visually during construction.

Wooden Template Used To Check Cone Clearance

43

We knew that Typhoon would enter a very hostile arena where every opponent would be trying to smash it to bits. We tried to imagine that Typhoon was made of balsa wood and cardboard and then considered what would happen if we applied a force in this direction or that direction. Which parts of the structure would take the maximum loads? Would it break or distort? This intuitive approach works and very little of our structure has failed although we have had some close calls and a lot of luck. Testing and particularly fights in the arena, have shown up some areas where we need to reinforce the structure or fixings.

Robot construction was undertaken at our cadet headquarters on normal parade evenings and also on Sunday afternoons over a 3 year period. We only had a bench with basic power tools and hand tools in our project room, so all of the machining and some precision work was undertaken in the garages of our instructors. The cadets were amazed at what they could construct with simple tools and a bit of ingenuity.

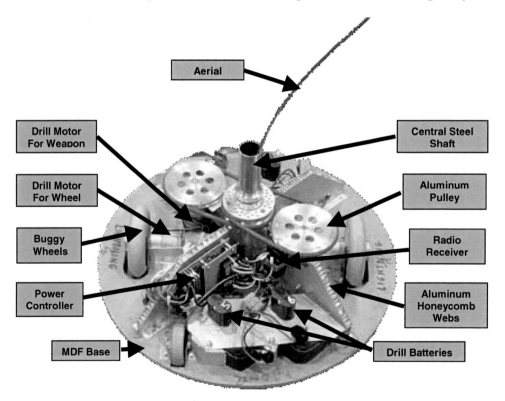

Construction of Typhoon Lightning

Chapter 7
Engaging and Avoiding the Enemy

Attack and Tactical Withdrawal

To avoid the enemy whilst spinning up to gain energy, our Typhoon family must be fast and agile. Our middleweight Typhoon used two 1 Horse Power wheelchair motors to drive independently a wheel at each side of the robot. This was more power than was needed as the wheels skidded under full power. The robot would turn on the spot with ease, but the motors were not perfectly matched for forward and reverse power which meant that Typhoon would follow a curved path in one direction. It thus required considerable skill to drive.

The most effective weapon is driver control and without it even the best robot will be beaten. A good driver can avoid the deadly blows from an axe and stop a flipper getting a face-on engagement. He or she must not only avoid the enemy but at the same time position his robot for an attack on the weakest part of the enemy. Attack is the best form of defence and if you are on the offensive your opponent will have little time to develop his own strategy and position himself to attack your weaknesses. The pit is a major hazard to Typhoon as every action has an equal and opposite reaction and every collision will send Typhoon skidding away in the opposite direction like the rebound of a snooker ball.

The judges give points for **Style, Control, Aggression** and **Damage** with damage getting the highest weighting. Typhoon glides around the arena with considerable style and when it positions for an attack exhibits considerable aggression. Unfortunately it needs time to replenish the energy in the weapon and must often make a tactical withdrawal and dodge its opponent for a few seconds. During this time its opponent can gain points for aggression. This is not generally a problem as we can cause significant damage and often deliver a knock out blow which immobilises him before the end of the contest.

Wheel Arrangement

Typhoon initially was fitted with casters fore and aft to balance the chassis, but during its first test on our parade square the casters broke and we therefore changed the design to small fixed wheels. The weight distribution on the chassis is fairly even so these stabilizing wheels supported very little weight and therefore skidded sideways easily when the chassis turned at zero or low speed. If a large straight line acceleration is demanded the front wheel comes off the ground and the weight on the rear wheel is increased significantly. This prevents sideways motion and helps to keep our robot straight during acceleration. Conversely, during deceleration the front wheel is loaded and this again assists in maintaining a straight course.

For Typhoon 2 we decided to use four wheels as we wished to keep the same power to weight ratio and we knew it would be easier to control in a straight line as the loaded drive wheels would need to skid sideways in order to turn. Another reason stemmed from our knowledge of aircraft systems where essential electrical and hydraulic systems are duplicated and isolated from each other so that in the event of battle damage the remaining system can get you home. The rules state that if a robot fails to move in a controlled manner for 30 seconds it is eliminated from the competition. It is therefore very important to maintain controlled horizontal movement even if the weapon is immobilized. Our design concept is therefore to use four Iskra 1 Horse Power motors to drive independently the four wheels. The front wheels are driven from one battery and the rear wheels from a second battery. This philosophy was proven in our 7th series semi-final when a drive chain came off and jammed one wheel. Gary noticed Typhoon 2 had a bias to the right, but it was still perfectly controllable and the audience and judges probably didn't even notice.

We could now drive Typhoon 2 easily in a straight line and even had to limit the power to the drive motors to prevent wheel spin. Unfortunately the friction between the drive wheels and the ground required full power to turn on the spot and we were using much more power to turn than we had allowed. After some deliberation and obtaining advice from the Tornado team, we decided to increase the motor to wheel drive ratio from 3 : 1 to 4.3 :1 by using a smaller drive sprocket and this allowed us to turn using much less power. The change had the side effect of reducing our top speed. But this speed was initially excessive and the reduction was deemed to be acceptable.

**Typhoon 2 Wheel Box Showing Independent Drive From Each Motor
To The Wheels Via Bicycle Chains**

Maximum Horizontal Speed.

Speed is Distance / Time. The maximum horizontal speed depends on the size of the wheels and how fast they can revolve. Knowing a few parameters it is easy to calculate the top speed of Typhoon and Typhoon 2. (They both use the same parts).

The maximum speed of each wheel motor is 6000 rpm or 100 revs per second.
The wheel radius is 0.08m and the wheels are driven by bicycle cogs and chain.
The motors are fitted with 7 tooth sprockets and the wheels have 30 tooth sprockets.

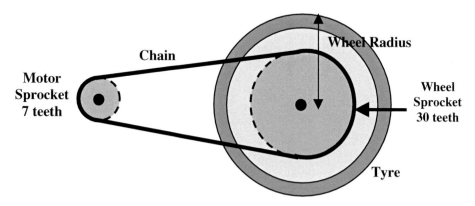

The motor to wheel drive ratio is the ratio of the teeth on the two sprockets

$$\text{Drive ratio} = 30 / 7$$
$$= 4.28 : 1$$

The motor therefore turns 4.28 revolutions to turn the wheel one revolution
The maximum wheel speed is the motor speed / drive ratio

$$\text{Max wheel speed} = 100 / 4.28 \text{ revs per sec}$$
$$= 23.4 \text{ rps}$$

The circumference of the wheel is 2π x radius
The maximum horizontal speed of the robot is the circumference of the wheel x its maximum speed of rotation

$$\text{Maximum horizontal speed} = 2\pi \text{ x } 0.08 \text{ x } 23.4$$
$$= 11.7 \text{ metres per second}$$
$$= \textbf{26 miles per hour}$$

In practice friction and other losses will probably limit the top speed to around 20 mph.

Maximum speed is not the most important parameter as the arena is not very big. More important is how much pushing power you have and how quickly you can accelerate.

Force and Acceleration

We apply linear force when we push an object along the ground.

Pushing power (linear force) and acceleration are connected by Newton's Second Law which states that if an object has an unbalanced force acting on it, it will accelerate in the direction of that force. Linear Force is the product of Mass and Acceleration $(\mathbf{F = ma})$ and is measured in Newtons (N).

Weight and Mass

The word weight is widely misused in everyday conversation. We often talk about the weight of a person or object being 70kg when what a physicist would mean is a mass of 70kg. The weight of an object is the gravitational force on the object. It depends on its mass and which planet it is on. The acceleration due to gravity on Earth is approximately 10 metres/sec^2 so a mass of 1 Kg actually has a weight (gravitational force) of 10 N.

Torque

Rotary Force is called Torque. It is the force applied to an object on an axle that causes the object to rotate about the axle. Torque is equal to the force applied to an object times the distance between the point of application and the centre of rotation and is measured in Newton Metres (Nm).

Linear Quantity	Symbol	Units	Rotary Quantity	Symbol	Units
Force	F	N	Torque	T	Nm

The Torque from one of our 1 HP drive motors when stalled is about 10 Nm. If we were to attach a bar 1 metre long to the motor shaft and let the other end push down on some bathroom scales the force on the scales would be 10 N.

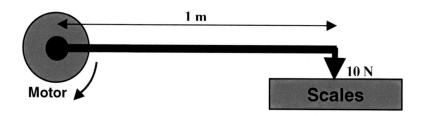

As Force = Mass x Acceleration and acceleration due to gravity is approximately 10 metres/sec^2 a force of 10 N is equivalent to a mass of 1Kg.

We can increase the force on the scales if we shorten the length of the bar. If the bar was half a metre long the force would be twice as much: 10 / 0.5 = 20 N.

Electric motors produce their maximum torque at zero rpm and the torque reduces as the motor speed increases. The Torque of our 1 HP electric wheel drive motors is 10 Newton Metres (Nm) at zero rpm (stalled) reducing approximately linearly to zero Nm at 6000 rpm.

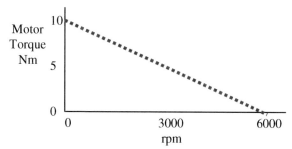

Motor Torque versus Motor rpm

We can increase the torque from our motor by gearing it down. Unfortunately the top speed will be reduced. This is why an electric screwdriver has a high torque, but rotates very slowly.

To calculate the torque at the wheels we multiply the motor torque by the drive ratio which is 4.28 : 1 for both Typhoon and Typhoon 2.

$$\text{Max wheel torque} = 10 \times 4.28$$
$$= 42.8 \text{ Nm}$$

The maximum force between each tyre and the ground which is available to propel the robot forward is then the wheel torque divided by the wheel radius

$$\text{Max Force Available} = 42.8 / 0.08$$
$$= \textbf{535N}$$

Friction

Without friction our robot will not move or even turn on the spot. The more friction we can get, the more force we can use for acceleration and pushing.

The coefficient of friction **μ** between the rubber tyres and the wooden arena floor is the horizontal force required to cause the robot tyres to just start to slip (i.e. the limiting frictional force) divided by the force of the robot acting downwards onto the floor.

Quantity	Symbol	Units
Coefficient of Friction	μ	None

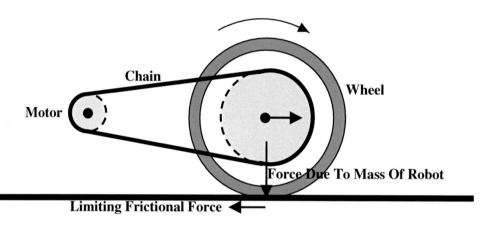

We locked the wheels and pulled 'Typhoon 2' horizontally with a spring balance. The robot started to slip with a pull indication of 90kg which is a pull force of 900N. The robot mass is 100kg and it therefore exerts a downward force on the floor of 1000N.

$$\mu = 900/1000$$
$$\mathbf{\mu = 0.9}$$

Knowing the coefficient of friction for two materials such as rubber and wood we can reverse this calculation to obtain the Limiting Frictional Force = **μ** x the force of the robot acting downwards.

$$\text{Limiting Frictional Force} = 0.9 \times 1000$$
$$\mathbf{= 900 \ N}$$

As there are four wheels sharing the weight equally, if the force from any one wheel exceeds a quarter of 900N that wheel will slip.

$$\text{Limiting Frictional Force (1 wheel)} = \mathbf{225N}$$

This means that at low rpm we are unable to use the maximum available wheel force of 535 N as the wheel will slip. The no-slip wheel speed can be calculated from the similar triangles in the diagram below:

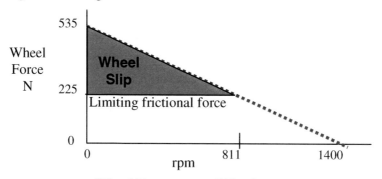

Wheel Force versus Wheel rpm

No-slip speed = 1400 x (535 – 225) / 535
= 811 rpm

To get maximum acceleration performance the driver must limit the torque manually by throttling back until the speed reaches 811 rpm when the robot will be travelling at 15 mph. He can then select full power.

Longitudinal Acceleration

Acceleration is the change in speed of an object in one second.

Linear Quantity	Symbol	Units
Distance	s	m
Velocity	v	m/s
Acceleration	a	m/s^2

Force is mass x acceleration; therefore Acceleration = Force/mass

Acceleration (from 4 wheels) = (225 x 4) /100
a = 9 metres per sec^2

To calculate the time to accelerate from rest (u=0) to 15 mph (6.7 metres per sec) we use the equation of motion **v = u + at**

As u=0
$t = v / a$
$t = 6.7 / 9$
t = 0.75 seconds

Wow! that's fast.

Distance Travelled

To calculate the distance travelled across the arena to achieve 15 mph in 0.75 seconds we use the equation of motion $s = ut + \frac{1}{2}at^2$

As u=0
$$s = \frac{1}{2}at^2$$
$$s = \frac{1}{2} \times 9 \times 0.75^2$$
$$\mathbf{s = 2.5 \text{ metres}}$$

In actual fact there will be some loss of torque in the chain drive and without a Formula 1 racing car type of traction control system the driver will be unable to adjust the throttle to achieve exactly the limiting frictional force throughout the acceleration. The practical result is therefore about twice these calculated figures. Nevertheless it's a very sporty performance.

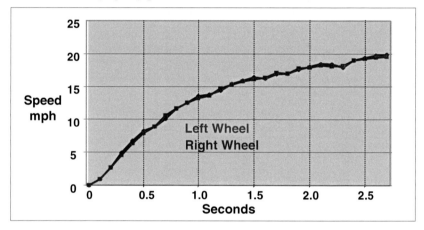

Actual Recorded Acceleration - Typhoon 2

Chain Strength

To calculate what strength of chain is needed to drive the wheels we recognise that the force (tension) in the chain will be limited by the wheel slipping. The maximum tension will correspond to the maximum force at the motor or wheel sprockets.

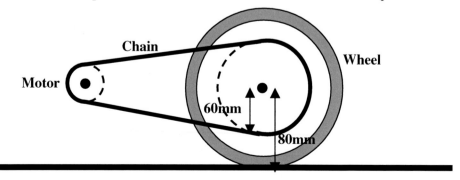

The normal maximum torque at the circumference of each of the four wheels is the Limiting Frictional Force x radius of the wheel

$$\text{Limiting Torque} = 225 \times 0.08$$
$$\textbf{=18 Nm}$$

To calculate the force (or tension) in the chain we divide by the radius of the big sprocket (0.06m)

$$\text{Force in chain} = 18 / 0.06$$
$$\textbf{= 300 N} \qquad \textbf{(The equivalent of 30 kg)}$$

The worst case is when all the weight of the robot is concentrated on one wheel (which means it won't slip) whilst that wheel motor is delivering maximum torque.

$$\text{Limiting Torque} = 535 \times 0.08$$
$$\textbf{= 42.8 Nm}$$
$$\text{Force in chain} = 42.8 / 0.06$$
$$\textbf{= 713 N} \qquad \textbf{(The equivalent of 71 kg)}$$

Good bicycle chain has a working load of about 2000 N and it breaks at a load of 9500 N (tensile strength). It is therefore able to carry our loads with ease.

Power

Power is the rate of gaining or using energy (doing work).

Electrical Power is measured in Watts (W) which is 1 Joule of energy (work done) per second.

$$\text{Power} = \text{Energy} / \text{Time}$$

Mechanical Power is often measured in Horse Power (HP)
A horse is capable of pulling twice the load of a very fit human.

$$1 \text{ HP is the same as 746 Watts}$$

A 1 HP (746Watt) electric motor will deliver 746 Joules of energy every second.

Quantity	Symbol	Units
Electrical Energy	E	Joules
Electrical Power	P	Watts
Mechanical Power	P	Watts or HP

Electric Wheel Motors

For ease of installation and servicing we decided to install our motors and wheels in two self contained power units that could be removed from the robot as separate units. For the Technogames Assault Course and Football competitions we did just that and transferred the two power units into the purpose built wooden chassis of Typhoon Rover. The 4 wheel independent drive proved very successful and Rover won the assault course competition with a new world record time.

Our 24 volt electric drive motors are rated at 1 HP (746 Watts). We use 24 volts as this is a standard voltage for Direct Current (DC) motors. If we had used 12 volt motors the current (Amps) for the same power would be twice as much and our wiring and motor speed controllers would have to be much bigger and heavier. If we used higher voltage motors the number of batteries would have been greater and this means greater expense and weight.

An electric motor changes electrical energy into kinetic (movement) energy. Our motors have permanent magnets and their speed is directly proportional to the voltage applied to the motor. The direction of rotation can be reversed by reversing the connections to the DC supply.

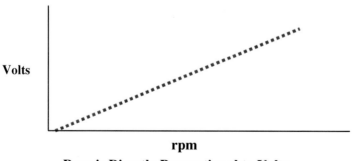

Volts

rpm

Rpm is Directly Proportional to Volts

We considered over-driving our 24 volt motors with 36 volts, but as the power we had was already causing the wheels to slip and we had more than enough top speed, it was unnecessary, would have generated much more heat and might have caused the motor windings to burn out.

Internal Heat

Electrical current causes heating. Motors, wiring and controllers will all get hot and waste power. The heat wasted is proportional to the square of the current x the resistance:

$$\text{Heat in Watts} = I^2 R$$
Heat increases as the square of the current

Permanent magnet DC motors produce their maximum torque at zero speed. Torque is proportional to the current flowing through the windings so these motors draw maximum current when stalled (approximately 200 Amps). This is a huge current and if the motor remained stalled for more than a second or two the windings would get red hot and burn out. When the motor is running the maximum current is about 35 amps and the torque is much less. The reduction in the current when the motor is turning is due to a phenomenon called **'back emf'** which is the voltage induced by movement of the magnetic field. This opposes the battery voltage and reduces the current flowing through the windings. The back emf is proportional to the speed of the motor.

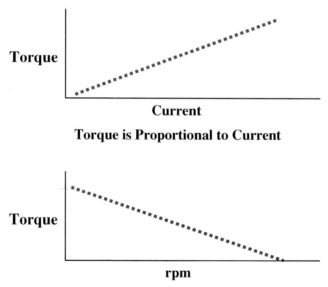

Current

Torque is Proportional to Current

rpm

Torque is Inversely Proportional to rpm

With four motors the theoretical combined stall current is 800 amps and the combined maximum running current is up to 140 amps. Fortunately the batteries, wiring and controllers all have resistance and this limits the stall current to about half this theoretical value. The maximum we have recorded during a fight is 432 amps and this caused our 24 volt sealed lead acid batteries to reduce their output voltage to 12 volts.

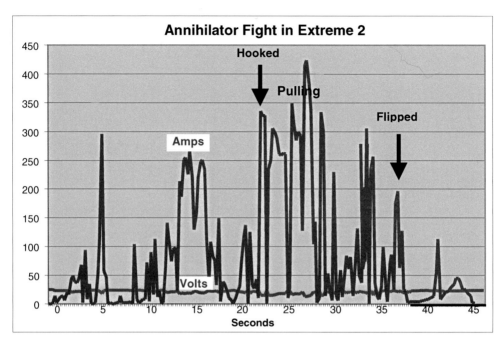

Steering Bias Due To Weapon Torque

We realised that when the weapon cone was spinning-up from zero, the chassis would want to rotate in the opposite direction and this torque would cause Typhoon to veer off to the right when we wanted to go in a straight line. To determine if this steering bias due to weapon torque would be a major problem we needed to calculate the torque caused by spinning up and compare this with the counter torque available from the wheels.

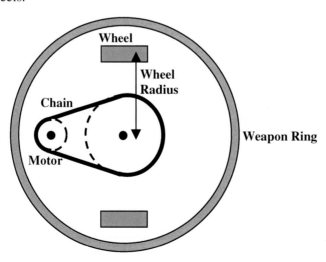

The relevant weapon parameters for our Middleweight Typhoon are:
Stall Torque of weapon motor = 14 Nm
Weapon Gearing = 3 : 1
Distance of wheels from centre = 0.22m

The torque due to the weapon spinning up is the motor torque x gear ratio.

Weapon Torque = 14 x 3
= 42 Nm

To calculate the horizontal rotational force of the chassis at the wheels due to the weapon torque we divide by the distance of the wheels from the centre.

Rotational force at Wheels = 42 / 0.22
= 190 N
= 95 N per wheel

Typhoon has two wheels, but is half the mass of Typhoon 2. The limiting frictional force from each wheel is therefore the same as Typhoon 2 = 225 N.

As 95N is less than half of 225N Typhoon should be controllable, but will initially have a strong bias to the right. It is therefore quite difficult to drive in the arena. The acceleration in a straight line will be noticeably less than maximum as some wheel torque has to be used to counter the weapon torque. As the weapon gains speed the motor torque reduces and the steering bias decreases considerably.

The four wheel arrangement of Typhoon 2 is not so susceptible to steering bias when spinning up and the consequential loss of acceleration is quite small.

Chapter 8
A Weapon Of Mass Destruction

Damage

The amount of damage we can do to an opponent is primarily how much **energy** we can hit him with. The more energy transferred in each strike, the greater the damage will be.

A single strike weapon, like an axe, which starts off stationary, and then is moved to strike the opponent can only gain energy in the short time between it starting to move and when it hits. A spinning disc or ring, however, can build up energy over a much longer period of time and release that energy instantly.

Energy

The energy of a body is a measure of the capacity which the body has to do work. To increase the energy of a body we have to do work on that body by applying a force.

Energy can exist in a number of different forms such as Gravitational Potential Energy and Heat, but the one we are particularly interested in is Kinetic Energy. This is the energy which Typhoon possesses due to its motion (either in a straight line across the arena to ram an opponent, or by spinning its weapon cone.

Energy cannot be created or destroyed, but can be changed from one form to another. Chemical Energy in the battery is changed to electrical energy and converted to kinetic energy in the motors.

Energy is measured in Joules, but because we can generate so much destructive energy we use Kilo Joules to avoid too many zeros. To understand how destructive an energy of 1 Kilo Joule is; Hypno-Disc in the 4th wars advertised its energy as 6 Kilo Joules. This energy, if it could be applied perfectly under a 100kg opponent, would toss him vertically upwards to a height of about 6 metres (The height of a single storey house).

We can verify this statement by calculating how much energy a 100kg robot would gain if it was dropped from a height of 6 metres onto the arena floor. This potential energy due to gravity is the Mass x Gravity x Height or **mgh**. (The acceleration due to gravity on earth is approximately 10 m/sec^2).

Potential Energy = 100 x 10 x 6
= 6 K Joules

Hypno-Disc has been hugely successful, but many of the modern robots competing in the 7th Wars greatly exceed this destructive energy. We will see later in this chapter by just how much Typhoon 2 exceeds the destructive energy of Hypno-Disc.

Linear Quantity	Symbol	Units	Rotary Quantity	Symbol	Units
Mass	m	kg	Moment of Inertia	I	kgm^2
Force	F	N	Torque	T	Nm
Distance	d	m	Angle	θ	radians (rad) or revolutions
Velocity	v	m/s	Angular Velocity	ω	rad/s or rpm
Acceleration	a	m/s^2	Angular Acceleration	α	rad/s^2
Potential Energy	E	joules			
Kinetic Energy	E	joules	Kinetic Energy	E	joules

Energy of the Typhoon Chassis

The Kinetic energy of the Typhoon chassis moving in a straight line across the arena is: ½ x mass x velocity squared

$$\text{Kinetic Energy of Chassis} = \tfrac{1}{2}mv^2$$

If the mass **m** is 100kg and the linear speed **v** of this 'Typhoon' battering ram is 20 mph (8.94 metres per second)   The kinetic energy of the Chassis is

$$\text{Kinetic Energy of Chassis} = 0.5 \times 100 \times 8.94^2$$
$$= 3996 \text{ Joules}$$
$$\sim \textbf{4 K Joules}$$

This is also the magnitude of the energy of a ramming robot such as Tornado or Storm II.

This is very dangerous but far less than the energy we want.

Energy of the Typhoon Weapon

The Kinetic Energy of the rotating Typhoon weapon is ½ x Moment of Inertia x angular velocity squared.

$$\textbf{Kinetic Energy of Rotation} = \tfrac{1}{2}I\omega^2.$$

The Moment of Inertia is the equivalent of mass in linear motion. It depends not only on mass, but on how far the mass is concentrated away from the centre of rotation. High inertia requires a large radius and a large mass concentrated at the edge.

To get maximum energy we need a high inertia I together with a high angular speed of rotation ω. This sounds simple. We make the weapon ring 1.2 metres in diameter (the maximum width allowed by the rules) and rotate it at the speed of light.

Size of the Weapon Ring

We had already decided that our design aim would be to make the rotating weapon cone about half the mass of the robot. As well as being the primary weapon, the outer ring is also the main defensive armour. For a given mass, the larger the radius of the weapon ring the thinner it must be and therefore more vulnerable to dents and distortion if hit by an enemy weapon. We decided 15mm thick steel would be the minimum thickness we could accept.

Mass Of The Weapon Ring

For Typhoon 2 we decided we needed a weapon ring with a mass of about 35kg which would allow about 15kg for claws, the cone and other structure needed to support it. It is fairly simple to calculate the mass of the ring if we consider it as two cylinders of slightly different radii. The mass of the ring is the mass of the outer cylinder minus the mass of the inner cylinder

Volume of a cylinder = Area of the circular top x the height 'h'

Volume of Cylinder = $\pi\,r^2\,h$

Mass of a cylinder 'm' = Volume x density 'ρ'

Mass of Cylinder = $\pi\,r^2\,h\,\rho$

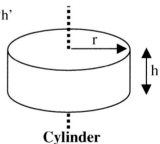

Cylinder

Mass of a ring is the mass of the outer cylinder – mass of the inner cylinder

$$\text{Mass of Ring} = \pi\,r_1^2\,h\,\rho - \pi\,r_2^2\,h\,\rho$$
$$= \pi\,h\,\rho\,(r_1^2 - r_2^2)$$

We tried several different combinations of radius, thickness and height and eventually settled on the following dimensions for Typhoon 2:

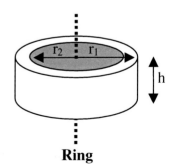

Ring

Outer radius of a ring	= 0.4m
Inner radius of ring	= 0.385m (i.e. 15mm thick)
Height of ring	= 0.12m
Density of steel	= 7850 kg/m^3

$$\text{Mass of Typhoon 2 Ring} = \pi \times 0.12 \times 7850 \,(0.4^2 - 0.385^2) \text{ kg}$$
$$= 34.8 \text{ kg}$$

This was very close to our desired mass and could be adjusted by small changes to the height and thickness of the ring.

The Speed Of Rotation

The speed of rotation has a large effect on the stored energy as energy is proportional to ω^2 (ω is the angular speed of rotation). It is measured in radians per second and there are 2π radians in a circle. To convert from revolutions per second to radians per second we multiply by 2π.

If you double the speed of rotation the energy increases by a factor of four.

At first sight it seems that we want to gear up our weapon motor to spin the weapon cone at the maximum speed we can achieve to give us the maximum energy possible. This is a false premise as there are several other factors that are vital to consider when making a practical full body spinning weapon that must cause maximum damage in a real fight. We believe the major factors are:

1.   The time to spin up to lethal speed.
2.   Mass Imbalance.
3.   Cutter Bite

Time to Spin Up to Lethal Speed.

It is no use having a weapon with a huge energy (of say 100 Kilo Joules) if it takes a minute to build up that energy. No opponent is going to sit back and let you spin up for that length of time.

If our motor could deliver 1 K Joule of energy every second the speed of rotation would increase as shown in the following graph. It will take a long time for the motor to achieve anything near its maximum no-load speed, but the weapon will have a reasonable destructive energy after only a few seconds.

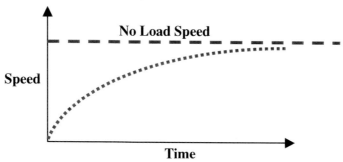

For our middleweight Typhoon the no-load speed of the weapon motor is 6000 rpm.

The reduction gearing = 6 to 1 (motor does 6 revolutions to turn weapon 1 revolution).

> Maximum Speed of Rotation = motor rpm / reduction gearing
> **= 1000 rpm**
> ω = 2 π rpm / 60 radians per second
> **ω = 105 radians per second**

We can alter the spin-up time (and maximum speed and stored energy) by changing the gear ratio.

If we change the gear ratio by a factor of 2 from 6:1 to 12:1 (motor does 12 revolutions to turn weapon 1 revolution) the maximum speed of rotation decreases by two = 500 rpm. This has two effects: The energy stored in the weapon reduces by a factor of $2^2 = 4$. The spin-up time also decreases by a factor of 4.

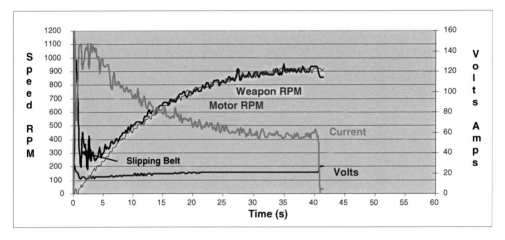

Actual Weapon Spin-up - Typhoon

The time to spin up the Typhoon 2 cone using our petrol engine is examined later in this chapter.

Mass Imbalance.

If the weapon cone rotates too fast any small imbalance will cause a major instability or wobble which in-extremis could cause the robot to self destruct.

A rotating mass tries to continue in a straight line and the centripetal force (towards the centre of the circle) holding it in place is **$mr\omega^2$** Newtons.

Force = mass x acceleration and is measured in Newtons. One Newton is that force which when acting on a mass of 1 kg produces an acceleration of 1 m/sec^2

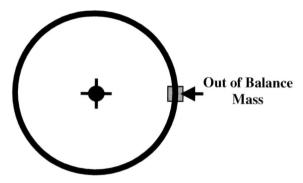

Out of Balance Mass

If we had an out of balance mass of 100 grams (0.1kg) at a radius of 0.5m rotating at 1000 rpm we can calculate its centripetal force.

$$F = m\,r\,\omega^2$$
$$F = 0.1 \times 0.5 \times (2\pi \times 1000 / 60\,)^2$$
$$\mathbf{F = 548\ N}$$

This is the equivalent of a weight of 54.8 kg and the same as if the out of balance mass had increased by a factor of 548. It is therefore very important that we balance our weapon. It is extremely difficult to balance a large weapon disk and if it gets a big hit it may distort slightly and cause a significant imbalance.

Fortunately the effect of an out of balance mass depends on how big a fraction it is of the complete rotating mass (actually its Moment of Inertia compared to the Moment of Inertia of the complete weapon cone). Nevertheless the out of balance force increases as the square of the speed of rotation and we must therefore keep the rotational speed down to a level where a small imbalance will not cause a major vibration or possibly a catastrophic disintegration of our robot.

The Cutter Bite.

This is how far each cutter tooth advances before the next tooth reaches the same angular position. It determines whether a weapon cuts or hits its opponent.

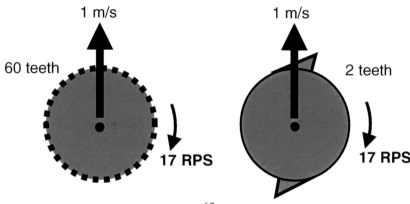

Consider a circular saw weapon with 60 teeth rotating at 1000 rpm (17 rps) and moving forward at 1 metre per second. The cutter bite is the speed forward times the time taken for each tooth to rotate to the same position as the previous tooth.

> Cutter bite = Speed forward x time for one revolution / number of teeth
> = 1 x (1/17) / 60
> **Cutter bite = 1 mm**

1 mm is insufficient for a tooth to catch on any part of the opponent and the disc will keep spinning and only slowly transfer kinetic energy to its target. The result is that the multi-tooth weapon will saw slowly into the bodywork of its opponent and its weapon energy will dissipate slowly causing little damage.

In contrast, a weapon the same diameter with only two teeth rotating at the same speed will advance **31mm** before the next tooth gets to the same angular position. A tooth will catch easily on the bodywork of its opponent and the disc will lose a lot of speed very quickly and may even stop. In doing so it will transfer a huge amount of its kinetic energy to its opponent in a millisecond, knocking it across the arena and causing major internal and external damage. This was the configuration we chose.

If the speed of rotation increases by a factor of two the cutter bite will halve.

Speed of Cutters

The maximum peripheral speed of the cutter tips is the circular distance travelled by the cutter in each second = the speed of rotation x the circumference.

> Max Peripheral speed = $1000 / 60 \times 2\pi \times 0.5$
> = 52 metres/sec
> **= 117 mph**

This is an interesting number to quote in an interview, but it does not have much practical significance.

Moment of Inertia

So how much energy is there in the heavy weapon of Typhoon 2?

To calculate the energy we first need to calculate the Moment of Inertia (M of I) of the weapon cone. Objects are reluctant to change their motion and if travelling in a straight line their inertia depends on mass. The Moment of Inertia for a rotating object is the equivalent of mass in linear motion. It depends not only on mass, but on how the mass is distributed within the shape of the object. Complicated shapes must be broken down into simple shapes. To calculate the total M of I we need to add up the M of I of each of its component parts.

1. The Steel Weapon Ring

The Moment of Inertia of a cylinder

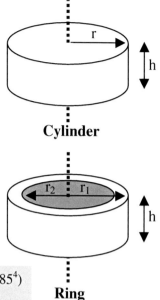

Cylinder

$$\text{M of I of Cylinder} = \frac{1}{2} m r^2$$
$$= \frac{1}{2} \pi r^2 h \rho r^2$$
$$= \frac{1}{2} \pi h \rho r^4$$

The M of I of a ring is the M of I of the outer cylinder minus the M of I of the inner cylinder.

$$\text{M of I of Ring} = \frac{1}{2} \pi h \rho r_1^4 - \frac{1}{2} \pi h \rho r_2^4$$
$$= \frac{1}{2} \pi h \rho (r_1^4 - r_2^4)$$

Ring

Inserting the values used previously

$$\text{M of I of Typhoon 2 Ring} = \frac{1}{2} \pi \times 0.12 \times 7850 (0.4^4 - 0.385^4)$$
$$= 5.37 \text{ kgm}^2$$

2. The Titanium Cone and its Supporting Structure

The cone of Typhoon 2 is a double skinned sandwich made of Titanium and Aluminum with an overall thickness of 30mm. The thickness of the outer armour is a military secret, but we can simplify the calculation by treating the cone as a thin aluminum cylinder with an equivalent height/thickness of 10 mm. (Density 2700 kg/m^3)

$$\text{The M of I of this thin cylinder} = \frac{1}{2} \pi h \rho r^4$$
$$= \frac{1}{2} \pi \times 0.01 \times 2700 \times 0.385^4$$
$$= 0.93 \text{ kgm}^2$$

3. The Hardox Steel Weapon Claws

Each of the four claws is approximately triangular in shape and their mass is about 1 kg each. This mass can be considered to act at the centre of gravity of each claw.

$$\text{The M of I of each claw} = \text{mass} \times \text{radius}^2$$
$$= 1 \times 0.45^2$$
$$= 0.2 \text{ kgm}^2$$

4. Total Moment of Inertia

Adding the component parts together we get the total Moment of Inertia of the weapon

$$\text{Total M of I of Typhoon 2 Weapon} = 5.37 + 0.93 + (4 \times 0.2)$$
$$= 7.1 \text{ kgm}^2$$

We can now calculate the maximum energy of the Typhoon 2 weapon:

$$\text{Maximum Energy} = \tfrac{1}{2}I\omega^2$$
$$= \tfrac{1}{2} \times 7.1 \times (2\pi \times 1000/60)^2$$
$$\mathbf{= 39\ K\ Joules}$$

This is 6½ times the destructive power of Hypno-Disc in the 4th Wars.

This maximum energy, if it could be applied perfectly under a 100kg opponent, would toss him vertically upwards to a height of about 39 metres. (That's through the roof of the studio hangar).

> ## Wow, we really have got an awesome, destructive weapon!

Time To Spin Up The Typhoon 2 Weapon Cone

The Petrol Engine of Typhoon 2 develops 5 HP (3730 Watts) at about 10,000 RPM, but only averages 1.7 HP (~1250 Watts) in the zero to 5000 RPM band (clutch slipping). It therefore transfers initially an energy of about 1250 Joules per second into the weapon.

In 5 seconds the weapon will have gained about 6 K Joules of energy. As the engine rpm increases the torque increases and it will gain energy at a faster rate and achieve 16 K Joules in about 10 seconds and 39 K Joules in around 20 seconds.

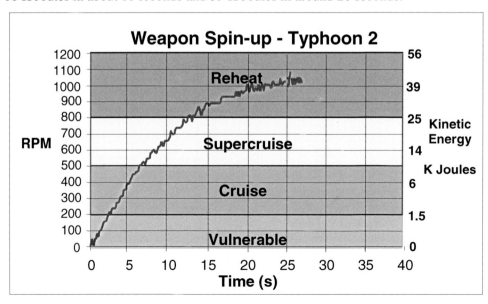

Graph Showing Time Taken To Spin Up The Typhoon 2 Weapon Cone

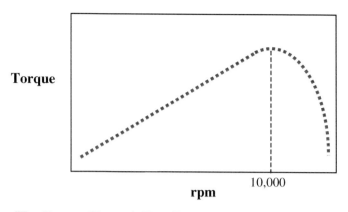

The Torque From A Two Stroke Petrol Engine Against RPM

A petrol engine is not the ideal power source to spin up the weapon cone quickly from zero. Like a Formula 1 racing car starting off from the grid you need high engine revs to get the power and torque required for a quick getaway. This means slipping the clutch when the engine revs are high and the weapon revs are very low. Our clutch is a chain saw centrifugal clutch which is automatic in operation. As the engine revs increase the three metal shoes are forced outward by centrifugal force and grip the clutch housing to transmit power.

Inside of Centrifugal Clutch Used on Typhoon 2

This is fine for the blade of a chain saw which has little inertia and accelerates to the same speed as the engine very quickly. In a chain saw the clutch becomes fully engaged in less than a second and has very little time to slip and heat up. When the saw is applied to the tree the clutch is fully engaged and transmits the full 5 HP into the saw blade. In contrast our weapon cone takes several seconds to reach a speed where the petrol engine can drive it directly (with no clutch slip) and transfer significant power.

In early qualifying fights we selected full throttle at the start and hoped the clutch would initially slip and then fully engage when the cone reached about one third speed. Unfortunately the slipping clutch heated up very quickly and continued slipping - getting hotter and hotter in the process. This caused it to slip more and our cone failed to reach a lethal speed before it was attacked and stopped.

We analysed the problem and recognised that two factors were inhibiting our performance:

1.   A very hot clutch.
2.   Dynamic (slipping) friction is much lower than static (engaged) friction.

Having recognised the problem we devised a technique to overcome it.   In subsequent fights we initially selected full throttle to get the cone moving, but when it reached about 200 rpm, we throttled back to match the engine speed to the speed of the weapon cone so that the clutch engaged fully and stopped slipping. We then slowly advanced the throttle again aiming to keep the clutch fully engaged during the subsequent acceleration with no slipping. This technique seemed to work well and as we gained confidence we were able to refine our technique and reduce even further the vulnerable time to spin-up.

A powerful electric motor would be a much better power source for our weapon as it has its maximum torque at zero revs and therefore does not need a clutch. Our middleweight Typhoon initially used a 1 HP electric motor for its weapon which was very successful. This was upgraded to a 3 HP electric motor for the 7th Wars which proved too powerful for our belt transmission. The rubber belt slipped, heated up and caught fire. We tried to be too clever and nearly lost the middleweight championship as a result.

We considered changing the petrol engine in Typhoon 2 for a powerful electric motor with a chain or toothed belt drive system. Unfortunately our weight and power analysis (described in more detail in chapter 8) has prevented us relying exclusively on electric propulsion for the weapon.

We are always looking for ways to improve the performance of Typhoon 2 and are currently assessing the possibility of harnessing the redundant power of our electric starter motor to help boost the acceleration of the weapon cone and reduce the time taken to spin-up to a lethal speed.

Replacing Energy

Our middleweight Typhoon has a maximum energy of 18 Kilo Joules, but due to frequent clashes averaged much less than this during its fights.

In our first fight (summarised in the table below) clashes occurred on average every 5.6 seconds and each clash expended an average energy of about 3.6 K Joules which had to be replaced by the weapon motor:

Secs	Target	Clash Severity	Approx Energy Expended	Time Int Secs
8	Zap	Small	3 K Joules	8
9	Genesis	Small	3 K Joules	1
11	Genesis	Medium	4 K Joules	2
15	Zap	Medium	4 K Joules	4
23	Zap	High	6 K Joules	8
29	Mammoth	Small	3 K Joules	6
33	Mammoth	Small	3 K Joules	4
37	Mammoth	Small	3 K Joules	4
43	Genesis	Small	3 K Joules	6
46	Hard Cheese	Small	3 K Joules	3
49	Genesis	Medium	4 K Joules	3
54	Hard Cheese	High	6 K Joules	5
73	Doom	Medium	5 K Joules	19
75	Mammoth	Very Small	1 K Joules	2
84	Doom	Small	3 K Joules	9
		Average	**3.6**	**5.6**
?	Matilda	High	5 K Joules	
?	Matilda	High	5 K Joules	
?	Refbot	Very High	7 K Joules	

If the weapon motor can deliver a continuous power of 1000 Watts (about $1\frac{1}{3}$ Horse Power) then it could transfer 1 Kilo Joule of energy to the weapon every second. The weapon would therefore gain an energy of 5 Kilo Joules every 5 seconds.

Real Destruction

We have examined the basic physical equations that control the energy and power of Typhoon 2, but there are other robot designs which have similar impressive specifications. Why is it that Typhoon 2 causes so much damage and destruction to its opponents and little damage to itself, whereas other robots with spinning weapon and equally high kinetic energy are not so effective or spectacular? This question is difficult to answer because it involves a number of diverse factors which contribute to the awesome destructive power of Typhoon 2.

What is it that causes damage?

Is it Force?

Linear Force (N) is the product of Mass and Acceleration ($\mathbf{F = ma}$). The internal damage is the force applied to components such as battery plates. When Typhoon hits an opponent the force of the collision causes the target to move. Each component (such as a battery) has mass and initially tries to remain stationary. The force causes each component to accelerate from zero to a particular velocity in a very short space of time and this causes vital components such as battery plates to bend and break.

Is it Acceleration / G Shock?

How quickly can you accelerate your opponent's chassis? How much g shock can you impart? Large accelerations and decelerations can be measured in G's. One G is the acceleration due to gravity which is approximately 10 metres per $\sec^2$. If you can accelerate an enemy robot by 100G every part in that robot will have its weight increased by a factor of 100 and a battery with a mass of 10kg will have an effective weight of 1000kg or a metric ton.

If you have a head-on car crash and are not wearing a seat belt you will hit the dashboard or windscreen, decelerate to zero extremely quickly and receive huge injuries. If you are strapped in and have an air-bag fitted these items will allow you to decelerate to zero over a longer period of time and you may be unhurt.

Is it Energy?

The energy of Typhoon is a measure of its capability to do work (J). This is a measure of its capability to apply Force over a distance (Nm) or its capability to accelerate a mass over a distance. High kinetic energy by itself does not cause destruction. It is how quickly it can be transferred to an opponent.

Is it Power?

Power is the rate of gaining or using energy (doing work). i.e. how quickly you can apply force over a distance. We say Typhoon 2 is a very powerful robot., but when not spinning it has power, but no kinetic energy and cannot cause much damage.

Is it Momentum?

Momentum is mass x velocity = mv. It is a vector quantity which means it has both magnitude and direction. The total momentum of a system remains constant. This means that if the Typhoon weapon hits an opponent and thereby loses momentum the opponent will gain the same quantity of momentum.

Is it Impulse / Impact?

An Impulse / Impact is a collision in which a large force acts for a small time. A hammer hitting a nail or a batsman hitting a ball. It is force x time of contact = Ft and is measured in N sec. The Impulse of a force is the change in momentum produced.

A Mathematical Analysis of Typhoon 2 Hitting an Opponent
For Highers and A Level students

In order to understand the basics of what happens when Typhoon attacks another robot, it is best to consider a very simple example, as illustrated below.

Before the collision let us assume that Typhoon's weapon ring is spinning at 1000 rpm (an angular velocity ω_1 of 105 (radians/sec). Let us also assume that Typhoon is moving from left to right with a linear velocity v_1 of 1 m/s (about 2 mph). Furthermore imagine that Typhoon consists simply of a chassis (non-rotating) with mass m_1 at its centre, and a spinning ring of radius r and mass m_2.

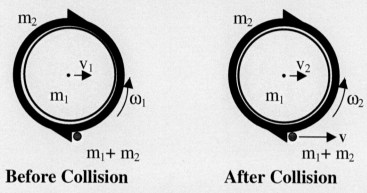

Before Collision   **After Collision**

The target is another robot. For simplicity, let's assume it has the same mass as Typhoon, i.e. $m_1 + m_2$, and that it consists simply of a point that is hit by one of the cutters. Although this is a gross simplification, it is realistic enough to explain what happens when the robots collide.

After the collision, let Typhoon's linear velocity be v_2 and its angular velocity ω_2; and let the velocity of the target robot be v.

So how can we work out what happens in the collision?

There are three very important principles of physics which we can use to help us. These are so-called *conservation principles*, which state that certain quantities are preserved unchanged before and after the collision. These principles are true provided there are no external influences on the mechanical system, and for the purposes of this investigation we shall assume this ideal case. In reality friction and damage to the robots will result in the results being somewhat different- we shall consider this later.

The three conservation principles we shall use are :

1. The principle of conservation of energy
2. The principle of conservation of linear momentum
3. The principle of conservation of angular momentum

Conservation of Energy

The first principle is perhaps the best known, and simply states that the energy of the system before and after the collision remains the same (provided it is an ideal, so-called *elastic* collision). In this case, the energy of interest is kinetic energy.

$$\text{Linear Kinetic Energy} = m\,v^2$$
$$\text{Angular Kinetic Energy} = \tfrac{1}{2} I\,\omega^2$$

I is the Moment of Inertia and for the simple case we need to consider here (a disk with all its mass at a radius r from a centre), the moment of inertia about that centre is given by $I = m\,r^2$.

$$E_A = \tfrac{1}{2}\,m\,r^2\,\omega^2 \qquad \text{for this type of disk}$$

So before the collision, the total energy of the system, E is

$$E = \tfrac{1}{2}(m_1 + m_2)\,v_1^2 + \tfrac{1}{2}\,m_2\,r^2\,\omega_1^2 \tag{1}$$

After the collision, the total energy is

$$E = \tfrac{1}{2}(m_1 + m_2)\,v_2^2 + \tfrac{1}{2}\,m_2\,r^2\,\omega_2^2 + \tfrac{1}{2}(m_1 + m_2)\,v^2 \tag{2}$$

Conservation of Linear Momentum

Linear momentum is the product of an object's mass and velocity.

$$M_L = m\,v$$

So before the collision, the total linear momentum of the system, M, is

$$M = (m_1 + m_2)\,v_1 \tag{3}$$

After the collision, the total linear momentum is

$$M = (m_1 + m_2)\,v_2 + (m_1 + m_2)\,v \tag{4}$$

Conservation of Angular Momentum

Angular momentum is the product of an object's moment of inertia and angular velocity. Angular momentum is conserved, but importantly only if it is measured about the *same* axis of rotation.

$$M_A = I\,\omega$$

So before the collision, the total angular momentum of the system, A, is

$$A = m_2\,r^2\,\omega_1 \tag{5}$$

After the collision, the total angular momentum is

$$A = m_2\,r^2\,\omega_2 + (m_1 + m_2)\,r\,V \tag{6}$$

The second quantity in this equation is most important - it is the angular momentum of the target robot *about the axis of Typhoon*. Since we have assumed that the target robot is simply a point with mass $(m_1 + m_2)$, its moment of inertia is simply $(m_1 + m_2)r^2$ and its angular velocity about Typhoon's axis is v/r.

We now have sufficient information to calculate what happens after the collision by solving the simultaneous equations (1) to (6) for the unknown values v_2, ω_2, and v.

From (1) and (2) we obtain
$$(m_1 + m_2)(v_1{}^2 - v_2{}^2) + m_2 r^2(\omega_1{}^2 - \omega_2{}^2) = (m_1 + m_2) v^2 \qquad (7)$$
From (3) and (4) we see that
$$v_1 - v_2 = v \qquad (8)$$
and from (5) and (6)
$$m_2 r(\omega_1 - \omega_2) = (m_1 + m_2) v \qquad (9)$$

Substituting from equations (8) and (9) for v_2 and ω_2 in equation (7) we see that
$$v = 2m_2(v_1 + r\omega_1) / (m_1 + 3m_2)$$
hence
$$v_2 = [(m_1 + m_2)v_1 - 2m_2 r\omega_1)] / (m_1 + 3m_2)$$
and
$$\omega_2 = [\omega_1(m_2 - m_1) - 2(m_1 + m_2)v_1 / r] / (m_1 + 3m_2)$$

So what does this all mean? Well, let's substitute some numbers into these equations, for various different masses m_1 and m_2. The maximum mass of a heavyweight robot in Robot Wars is 100kg, so $(m_1 + m_2)$ is 100kg. To make it even simpler, make m_1 and m_2 the same, that is 50kg. In fact this is about the right value for Typhoon 2.

Calculating the results, we find that the change of momentum for the target robot is 2150 kgms^{-1}, and the value for theTyphoon 2 chassis is precisely half this value. The force acting on the two robots is propotional to the change of momentum, so twice the force is applied to the target robot as to Typhoon's chassis. If we could change the mass distribution to make the weapon ring heavier and the chassis lighter, even more force could be applied to the target and the weapon would be even more devastating.

Chassis m1 (kg)	Ring m2 (kg)	Horiz Vel v1 (m/s)	Ang Vel w1 (rad/s)	Target V (m/s)	Energy on Target (KJ)	Energy on Chassis (KJ)	Energy Distrib Ratio
10	90	1	105	27.6	38	4	10.8
20	80	1	105	26.5	35	6	5.4
30	70	1	105	25.1	31	9	3.6
40	60	1	105	23.5	28	10	2.7
50	50	1	105	21.5	23	11	2.2
60	40	1	105	19.1	18	10	1.9
70	30	1	105	16.1	13	8	1.6
80	20	1	105	12.3	8	5	1.5
90	10	1	105	7.2	3	2	1.5

Note that a robot with a 20 kg weapon only imparts 1/3 of Typhoon 2's energy on its target and receives a much larger proportion of its energy back into its own chassis.

The most important factor is the amount of energy imparted to the target in the collision, because in the real world this energy is what does damage. In the ideal world energy is conserved, and in a sense in the real world it is - for a fraction of a second before the energy is dissipated and causes damage. In the collision analysed above, the target robot gains 23 K Joules of energy (about 60% of the maximum energy available). Typhoon 2's chassis only acquires about ½ of this amount. If we could make the weapon ring heavier and the chassis lighter, we could impart more energy to the target and less to our own chassis. Unfortunately, the need for drive motors, batteries, engine etc. is such that it would be very difficult to improve significantly on Typhoon's 50/50 mass distribution.

It is this 2:1 ratio, and one other factor – mass distribution, that allows Typhoon to carry out devastating attacks with minimal risk of internal damage. Basically the mass distribution of Typhoon ensures that the target always receives most of the energy in the attack, whilst the weapon claws ensure that the energy is concentrated on the smallest possible target area to cause the most damage.

**The Remains of Hammer Head 2 After Meeting Typhoon 2
in the Heats Semi-Final**

The reaction forces acting on Typhoon's chassis are well distributed by the main rotation bearings and the bumper ring between the inside of the weapon ring and the chassis. This large interacting area, together with strategically placed shock absorbers, minimises the risk of Typhoon damaging itself.

The Typhoon 2 weapon ring is unlikely to stop rotating after a single collision as the cutter bite will decrease and then disengage as the cone continues to rotate after the initial impact and the two robots ricochet off in opposite directions.

The opponent is not hit at its centre of gravity so every collision will impart a rotation as well as a linear acceleration to its chassis. The acceleration and damage will be greatest at the corner receiving the hit and decrease towards the centre of the rotation. The degree of rotation and horizontal movement experienced by both Typhoon and its target will also depend on the friction between the tyres and the arena floor.

Typhoon 2's Hardox Claws Hit the Titanium Shell of its Opponent

Aim For a Kill - Typhoon 2 Takes No Prisoners

Chapter 9
We Need a Lightweight Nuclear Power Source

Each fight lasts 5 minutes and we must still be moving and fighting at the end. The power source must be capable of supplying the full running current to the motors and also have the capability of supplying the peak currents when we need maximum power to get us out of trouble. How much power do we need to drive the wheels and how much for the weapon?

Power Required By Wheels

The electrical power required by the wheel motors is very difficult to estimate as every fight is different. There will be periods when we are stationary for short periods pushing against another robot with motors stalled. At another time we will be dodging and weaving round the arena avoiding an attack whilst the weapon gains energy. At other times we will be positioning for an attack and turning fast at slow speed. Our data logging indicates that with four 1 horse power motors driving the wheels, we need a power source which will deliver up to 100 amps for 5 minutes

Battery capacity required = 100 x 5 / 60
= 8 Ah

The Amp-Hour rating of a battery specifies its capacity to store energy. It is the number of amps the battery should be able to supply over a one hour period. Most battery manufacturers specify the capacity of their batteries when supplying only a few amps over a long period of time. If we demand very high currents from a battery it will heat up due to its internal resistance and cell chemistry (heat is proportional to current squared). This degrades the efficiency of the battery and it might only supply about half the rated capacity. To supply 8Ah we therefore need a much larger battery with up to twice the energy capacity (16 Ah).

Battery capacity can also be measured in Joules = Watts x time, or Amps x volts x seconds. A 24 volt 8 Ah battery would contain:

Capacity = 24 x 8 x 60 x 60 Joules
= 691 K Joules

We can calculate the heat energy wasted in the battery by measuring the volts dropped across the terminals when it is delivering a high current.
If the internal resistance of the battery is 0.01 ohms and it is supplying 320 amps

Volts dropped = Current x Resistance   **V = IR**
Volts dropped = 320 x 0.01
= 3.2 volts
Watts of heat = Current x Volts dropped   W = I

V

= 320 x 3.2

This is about the same as a single bar electric fire.

Power Required By The Weapon

The weapon motor will be going continuously throughout the fight and will have to be reasonably powerful to spin up the weapon cone quickly. If we assume we need to achieve an energy of 8 K Joules in 4 seconds we must transfer 2000 Joules from the motor to the weapon every second. This will require a motor which can deliver a power of 2000 watts. Assuming an efficiency of 65%.

$$\textbf{Power required} = 2000 / 0.65 \text{ watts}$$
$$= 3000 \text{ watts}$$
$$\textbf{or 4 HP}$$

The power required by the weapon is therefore very similar to that required by four 1 HP wheel motors and a similar battery capable of supplying 8Ah is required. (i.e. A sealed lead acid battery specified by its manufacturer as 16 Ah)

This battery we have already seen contains 691 K Joules of usable energy. Delivering 2 K Joules per second the battery would last for 345 seconds or 5.75 minutes.

Weight Analysis

Four 1 HP Iskra wheel motors	9 kg
Two 16 Ah 12 volt sealed lead acid batteries	18 kg
Dual speed controller	1 kg.
Total mass of wheel drive system is approximately	**28 kg.**
One 3000 Watt (4 HP) Lynch motor	11 kg.
Two 16 Ah 12 volt sealed lead acid batteries	18 kg
Speed controller	1 kg.
Total mass of weapon drive system is approximately	**30 kg.**
Mass of both Drive and Weapon power systems	58 kg.
We want to fit a weapon with a mass of	50 kg
We must allow for a chassis and supporting structure	10 kg minimum
Total Mass of Robot	**118 kg**

which is 18kg over the 100kg limit.

What can we do?

We were not rocket scientists, but we could see that the sealed lead acid batteries were the biggest contribution to our weight problem (36kg). Another cadet brain storming session came up with three ideas. The first idea from the cadets was to fit a nuclear power pack which we had to discard on safety grounds. This left us with two choices: We could fit lighter batteries or we could install a petrol engine for the weapon. We knew that petrol engines were unreliable to start, were difficult to control and install, and had a habit of cutting out in the arena. As far as we knew, no petrol engine robot had got very far in a Robot Wars competition. Nevertheless we decided eventually, after much discussion and analysis, to fit lighter batteries for the wheel motors and a petrol engine to power the weapon. Unlike batteries, which heat up and lose efficiency as the fight progresses, a petrol engine can deliver the same power to the weapon at the end of a fight as it does at the beginning.

Petrol Engine Power Source

A suitable 5 HP two stroke chain-saw engine was identified and this weighed 5kg. We chose a chain-saw engine as these have carburetors that can operate in any orientation. It would need a small amount of petrol, but petrol can provide about 100 times the energy of an equivalent weight of batteries. As a bonus, on examining the rules we found that consumables such as petrol were not included in the weight limit. With a petrol tank plus servos to control the throttle and choke we were still below 6kg and our excess weight was solved. We even had enough spare weight to install an electric starter motor for the engine which overcame our previous worry about the engine cutting out in the arena.

5 HP Chain Saw Engine of Typhoon 2

A starter needs a freewheel like a bicycle to avoid the engine back-driving the electric starter motor at high speed. They are normally reliable, but when we arrived at the 7th wars we found that the freewheel was freewheeling in both directions and we couldn't start the engine. When the panic had subsided we squirted some lubricant into the assembly and this seemed to cure the problem for a period of about 30 minutes. This freewheel was to provide us with some real heart stopping moments as we progressed through the 7th Wars competition.

Starter Freewheel (Our Achilles Heel)

We knew that a two stroke engine only develops significant power and torque at high rpm. We would definitely need a clutch and this would undoubtedly slip and get very hot.

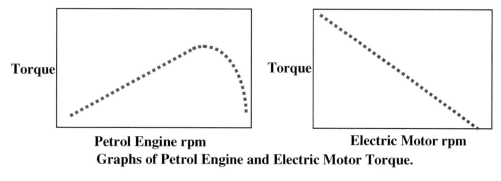

Petrol Engine rpm   **Electric Motor rpm**
Graphs of Petrol Engine and Electric Motor Torque.

We debated whether to fit a fire extinguisher in case the fuel caught fire, but in the end we did not have enough weight margin to fit one and the decision was made for us.

Fuel Capacity

How much fuel would we need? The rules allowed a maximum of 500 cc so we hoped that this would be enough for 5 minutes at a continuous full throttle. Petrol engines normally consume 0.3kg of fuel per Horse Power per hour. Our 5 HP engine would therefore consume 1.5kg of fuel per hour at full throttle and we had to last 7 minutes (5 minutes fighting plus 2 minutes at idle when entering the arena).

$$\text{Fuel required} = 1.5 \times 7 / 60 \text{ kg}$$
$$= 0.0175 \text{ kg}$$

Fuel has a specific gravity of about 0.75 so we would need a volume of

$$0.0175 / 0.75$$
$$=233 \text{ cc}$$

We therefore sized our fuel tank to contain 400cc of petrol as we did not want it too full in case we were inverted and we wanted the fuel feed pipe to be covered in fuel at the end of the fight. In the competition we discovered that we used about 200cc during a 5 minute fight and our calculation was reasonably accurate.

Petrol Tank Showing Its Protective Shield In Case We Get Stuck on the Flame Pit

Battery Technology

Batteries are the heart and blood of our robot and we took great care of them to avoid them letting us down. Cost dictated that we initially used Sealed Lead Acid (SLA) batteries for the wheel motors of Typhoon and Typhoon Rover and had to accommodate their considerable weight. When we had won the middleweight championship two years running and had a proven track record, we were able to approach a battery manufacturer and obtain sponsorship of SAFT Nickel Metal Hydride (NiMH) batteries for Typhoon 2. These are very rugged and have a very high energy density. The maximum continuous current is 40 amps and 100 amps can be demanded for a few seconds. The voltage of the cells also stays constant until they are almost fully discharged. Unlike the Nickel Cadmium (NiCad) batteries we use in the lightweight and featherweight Typhoons they do not have a nasty characteristic called 'memory effect' which means you have to discharge them fully before you can recharge them again to full capacity.

Nickel Metal Hydride (NiMH) batteries only came in relatively small 1.2 volt cell sizes and the maximum capacity for high powered 'C' size cells was about 3 Ah. To increase the voltage we add cells together in series. To increase the current capacity we add battery packs with equal voltage and current capacity together in parallel. To minimise the resistance of the connections between individual cells, we had low resistance tabs welded to the top and bottom of each cell.

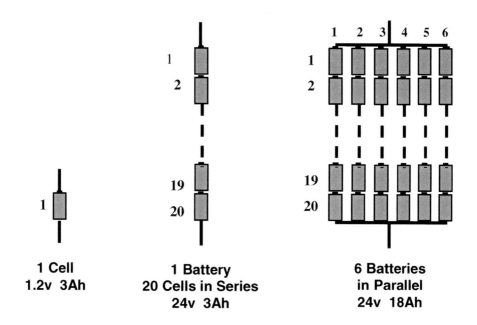

1 Cell
1.2v 3Ah

1 Battery
20 Cells in Series
24v 3Ah

6 Batteries
in Parallel
24v 18Ah

We needed a 24 volt Nickel Metal Hydride battery with a capacity of at least 16 Ah Each cell is 1.2 v, has a capacity of 3 Ah and a mass of 60 grams.

Number of cells in series in each battery = 24 / 1.2
= 20 cells
No of batteries in parallel = 16 / 3
= 6 batteries
Total number of cells = 120 cells

This is a very large number of cells and as we required at least one extra set of spare batteries charging up for the next fight we needed at least 240 cells.

Mass of batteries = 120 x 60 / 1000
Mass of batteries = 7.2 kg

This is a huge 10.8kg saving in mass compared with Sealed Lead Acid (SLA) batteries of the same capacity.

Charging is another problem with that number of cells. Most chargers only charge NiMH cells at low current which would take too long if we had to do several fights in one day. High power chargers are expensive and heavy and difficult to get for 24 volt batteries. We were also advised that the individual 24 volt batteries should be charged separately to ensure that they all received a full charge. Neil therefore designed a purpose made charger with eight separate outputs. The most reliable way to determine when a battery pack was charged was to monitor the temperature and switch to trickle charging when the battery pack temperature reaches 40 degrees centigrade. We therefore buried thermistors (devices whose resistance changes with temperature) in the centre of each battery pack.

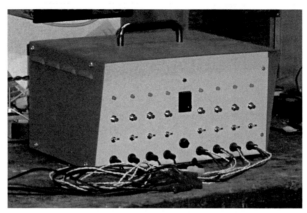

Purpose Built Battery Charger

Electrical Components and Connections

Balancing the electrics was important, batteries, motors, speed controllers and wiring all had to be matched to each other. Over-rating or under-rating one or more of these items can lead to inefficient use of materials, excess weight, increased costs and over stressing of key components.

Wiring

The size of the wire joining the batteries to the motor is a very important factor. Our middleweight robot 'Typhoon' had a bash which shorted the wires together inside the motor. The resultant current drain caused the thick wires to melt and we ended up with several gaps in the insulation of our wiring looms.

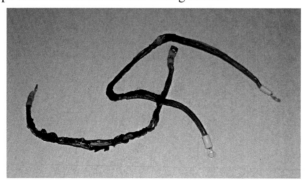

Melted Wiring Insulation

The thicker the wire the higher is its current handling capacity and the lower its resistance. Household mains wiring is normally sized to carry 13 Amps. With currents of 400 Amps plus travelling through the wires of Typhoon 2, much thicker cable must be used, as even a tiny resistance of a few milli-ohms will cause considerable heat.

The resistance of a wire depends on the material it is made of, its length, its thickness and its temperature (The resistance increases as it gets hotter).

For the 7th wars our high current wiring was upgraded to dual wall ETFE type. Although stiff and difficult to work with it can withstand very high temperatures. It is also thin walled and very lightweight for a given current rating.

If the resistance of the wire is 0.01 ohms

$$\text{Volts dropped} = \text{Current x Resistance} \qquad \mathbf{V = IR}$$
$$V = 400 \times 0.01$$
$$\mathbf{V = 4\ volts}$$
$$\text{Watts} = \text{Current x Volts dropped} \qquad \mathbf{W = IV}$$
$$= 400 \times 4$$
$$\mathbf{= 1600\ Watts\ of\ heat}$$

The Removable Link

Connectors like the screwed plastic joiners you buy for mains electrics are useless and will soon burn out. Heavy duty eyelets soldered or preferably crimped onto the wire are much better. The rules require a single electrical connector that can be removed physically from the robot. This safety device is known as the 'Removable Link' and when it is detached from the robot all the electrical circuits are isolated and rendered inoperative. This was a big challenge as this plug link carries the full current from all the batteries. A mains electrical plug and socket is only rated at 13 Amps and would arc, heat up and burn out.

We knew that the surface area of electrical contact was important to minimise resistance, so on our middleweight 'Typhoon' we made a special copper disc that screwed onto another disc which was cut into two parts. When the two discs were in contact power flowed from one side to the other through a large surface area of copper.

Home Made Removable Link At Base of Central Shaft

This home made connector worked, but showed some evidence of arcing, so when we were offered a high-power professional connector to replace it we jumped at the offer. Our robots all now have special low resistance high-power connectors. The lightweight robots are rated at 100 Amps, the middleweight is rated at 200 Amps and the heavyweight is rated at 1000 Amps

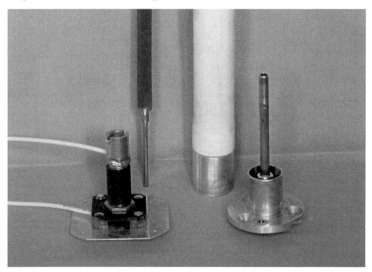

High Power Connectors Modified to be Inserted Down Central Shaft

Chapter 10
How Can We Tame This Mechanical Beast

Radio Control

Radio Control is the obvious way of controlling a robot. We purchased a suitable model aircraft type set from our local model shop which was modified to use the 40 MHz band which is for ground vehicles. This was a 6 channel Futaba 6X digital proportional R/C system. The transmitter worked by modulating the frequency about a central datum frequency. The Receiver decodes this Frequency Modulation (FM) and converts it into a series of pulses with varying width.

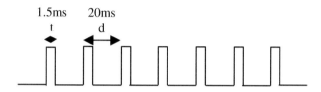

Pulse Width Modulated (PWM) Signal From Receiver to Servo

The pulses are repeated every 20 milli seconds and the width (t) of the pulse determines the position of the servo. A pulse of 1.5 ms is the centre position and pulse widths of 1.0 ms and 2.0 ms are the two extremes.

FM is a well known and fairly reliable method of radio control, but we knew that there would be all sorts of extraneous signals floating around the arena from arc lights and other radios, plus the arcing from the brushes of our own motors. We therefore decided to pay a little more and buy a transmitter and receiver that operates by coded FM pulses. With this system the transmitter converts each joystick, switch, knob and button position into a 10 bit digital word, plus extra bits to enable the receiver to verify the word. This Pulse Code Modulation (PCM) is more reliable in areas where there is a lot of electrical interference.

At near maximum distance between transmitter and receiver, or if there is a lot of interference, the receiver detects an error in the transmitted pulse coded data, ignores this bad data and uses the last known good data. The first effect is that the robot appears to have a small delay in responding to a new command. If the interference gets worse and valid data is not received for half a second, a fail safe device operates and all the six channels go to pre selected values which causes motors and weapons to go to a zero speed safe condition. This Failsafe function is mandatory.

Electric motors emit sparks in operation and these sparks cause electrical interference. We therefore connected small capacitors across the motor terminals and from the terminals to earth, as close as possible to the motor. This helped to suppress the electrical spikes.

Aerial

The aerial is a very important part of the receiver and must be a particular length to get maximum sensitivity. Light travels at 300,000 Kilometres per second and the frequency is 40 Mega Hertz.

$$\text{Wavelength} = \text{Speed} / \text{Frequency}$$
$$= 300,000,000 / 40,000,000$$
$$= 7.5 \text{ m}$$

The aerial length should ideally be this wavelength. But it would not be practical to carry an aerial this long so a quarter wavelength aerial is used = 1.9 m long. This is still too much wire for the arena, but fortunately the receiver manufacturer reduces it to about 1m by fitting an inductor near the bottom to compensate for the extra length which is missing.

As much of the aerial as possible should be outside the metal shell of the robot or the metal will shield the aerial from the radio waves and the robot may go absent without leave (AWOL) in the arena. Our metal cone is particularly good at shielding radio waves so we brought the aerial up the central shaft within a plastic tube and at least half if it is free of any shielding whatsoever. It is also in a place where it is unlikely to get damaged unless the robot is flipped over and if this happens, damage to the aerial is the least of our worries.

Motor Speed Controllers

Our middleweight 'Typhoon' uses wheel chair motor speed controllers. These are normally controlled by the disabled person operating a rotary potentiometer (variable resistance). Initially the radio receiver drove a standard servo which mechanically moved the potentiometer to select forward and reverse. Later, our electronics wizard Neil made a special interface which plugged directly into the receiver and ran the speed controllers without any mechanical components.

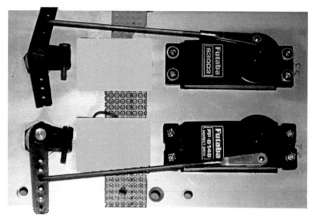

Servo Linkage Used In Middleweight Typhoon

A motor speed controller works by varying the average voltage sent to the motor. The most efficient way to do this is to switch the full 24 volt supply voltage on and off again very quickly in a succession of pulses. If the switch is on for the same amount of time that it is off the motor will see an average of 12 volts and run at half speed. If the switch is on for longer than it is off the motor will see a higher average voltage and run faster.

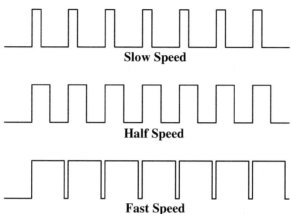

Slow Speed

Half Speed

Fast Speed

A motor speed controller takes the signal representing the demanded speed and drives a motor at that speed in the correct direction. High power motor speed controllers are very complicated as they not only have to vary the current to the motor, but also have to swap over the two motor connections whenever reverse is selected.

Ordinary transistors can be used to switch these high power pulses on and off very quickly and are adequate for small motors drawing up to 20 amps. For bigger motors drawing 200 amps or more, devices known as MOSFETS (Metal Oxide Semiconductor Field Effect Transistors) are much better. These are devices that can turn very large currents on and off under the control of a low signal level voltage. For Typhoon 2 we used Vantec speed controllers which use 48 MOSFETS rated at 70 Amps each to drive two motors in forward and reverse.

Even MOSFETS have a small resistance and therefore heat up when controlling several hundred amps of current.

$$\text{Heat in Watts} = 200^2 \times 0.001 \qquad \mathbf{I^2 R}$$
$$= \mathbf{40\ Watts}$$

It is very important not to let devices such as transistors or MOSFETS reach too high a temperature, so they are mounted on aluminium or copper heat sinks to conduct the heat away from the chip and allow it to dissipate into the air.

To drive a motor a minimum of four MOSFETS $Q_1 Q_2 Q_3 Q_4$ are arranged in what is called a full bridge circuit. To control very high currents each individual MOSFET can be replaced with several MOSFETS connected in parallel to increase the current handling capacity.

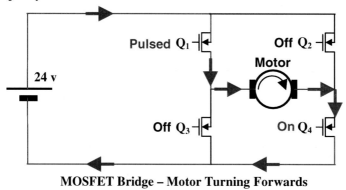

MOSFET Bridge – Motor Turning Forwards

To make the motor go forwards Q_4 is turned on and Q_1 is pulsed on and off. The current flows through the motor from left to right. To make the motor go backwards Q_3 is turned on and Q_2 is pulsed on and off. The current then reverses direction through the motor and it turns in reverse.

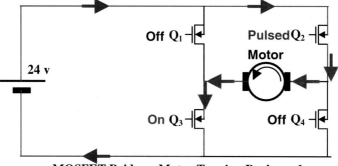

MOSFET Bridge – Motor Turning Backwards

On / Off Switch

To control circuits such as the starter motor and the engine kill relay the pulse width of the receiver output is sampled and compared with a preset value. The motor or relay is switched on or off depending on whether the pulse width is greater or less than this preset value.

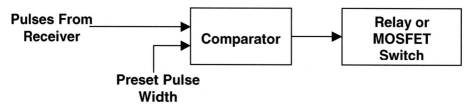

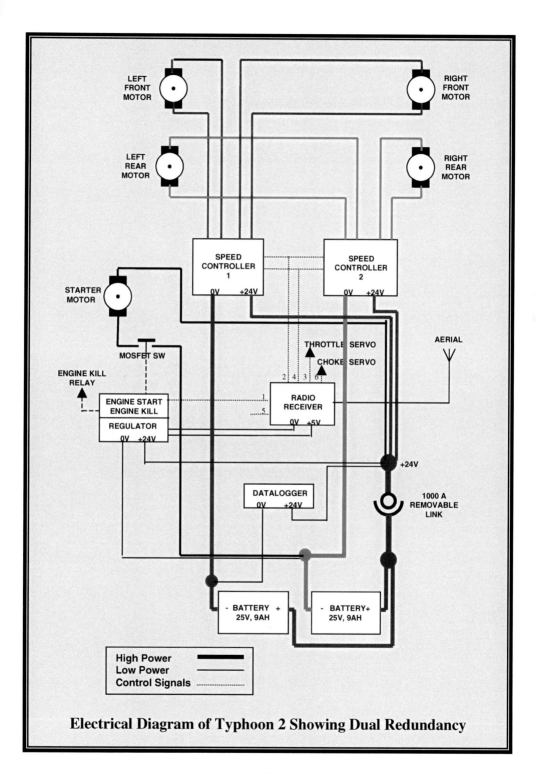

Electrical Diagram of Typhoon 2 Showing Dual Redundancy

Control Laws

Our Typhoon robots employ differential, 'tank type' or skid steering. This means that they are turned by varying the speed of wheel rotation on each side of the robot. If the left wheel is going forward and the right wheel is in reverse the robot will spin on the spot. (This is known as a zero turning circle). Moving one control lever on the transmitter forward commands both motors to run forward at the same speed. Moving the same (or another) lever sideways causes one motor to go forward and one to go in reverse. If selections are made at the same time the robot will turn whilst moving forward or in reverse. The lateral demand is added to the longitudinal demand to provide the required mixed command to each motor. The robot then turns in the direction of the slower wheel.

To provide precise control at slow speed for accurate manoeuvring near the pit it is useful to have the control sticks more sensitive around the central position. We were able to achieve this on our transmitter by selecting a facility called 'exponential control laws'. This allows the sensitivity around zero to be adjusted according to an exponential curve.

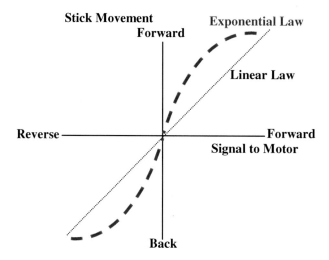

A gyro is a very useful device to help you drive in a straight line. This should have a limited authority so that you can override its authority if necessary. We were given a solid state silicon gyro from BAE Systems and fitted it on Typhoon. For Typhoon 2 (same wheel boxes as Typhoon Rover) the four wheels provided enough stability and no gyro was needed. The controllability of Typhoon Rover in the Technogames assault course was particularly commented on by the experts.

As well as controlling the wheel motors we also wanted to control the weapon motor. This was electric on our Middleweight Typhoon and we thought a simple on-off relay would suffice. We installed one rated at 180 amps, but it burned out during our first fight and was replaced by one of Neil's special solid state MOSFET switches.

We fitted small electrically initiated theatrical smoke cartridges to Typhoon for its first fight. These were controlled by a standard servo and rotary selector switch. As we demanded more servo the switch rotated and made contact with each cartridge in turn. This worked well, but there was so much smoke in the arena from the arena hazards that our countermeasures were hardly noticed. The smoke also covered the chassis and electrics in black grunge which was difficult to remove.

Weapon Control Box

We wanted the driver to concentrate on his driving and not be distracted by the Weapon Systems Operators (WSO) leaning over to operate switches etc. We therefore modified the rear plug on the transmitter to allow a separate weapon control box to be plugged in. This was very successful and enabled us to fit several different controls that have different functions in each of our robots.

 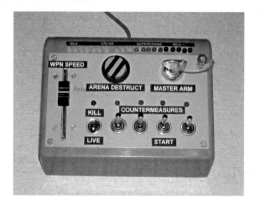

Transmitter   **Weapon Control Box**

The box was made by Hazel, Graeme and Alistair, our WSO's, and as well as a 'Master Arm' key it featured a large black and yellow striped button labelled 'SELF DESTRUCT' which set off all the smoke cartridges at once. For Typhoon 2 this button started the petrol engine. After the 7th Wars semi final (when we destroyed the side of the arena) the button was relabelled 'ARENA DESTRUCT'.

The transmitter controls of Typhoon 2 were configured as follows:

Channel 1	Weapon Box Switch 1	Engine Kill
	Weapon Box Switches 2-4	Reserved for Countermeasures (Smoke)
	Weapon Box Push Button	Starter Motor
Channel 2	Right Stick Fwd/Back	Fwd / Rev
Channel 3	Weapon Box Slider	Petrol Engine Throttle
Channel 4	Left Stick Left/Right	Left / Right
Channel 5	Right Top Switch on Tx	Reserved For Self Righting Mechanism
Channel 6	Flap rotary knob	Choke
Dual Rate	Right D/R Switch	Control Law Gain Change

The reason that the forward/back command is on the right stick and the left/right control is on the left stick is because our primary driver likes this configuration. Being a pilot I personally prefer both control axes to be on the same stick.

Driver Training

870 (Dreghorn) Squadron has 80 cadets of which 40% are girls. We held a driving competition for them over several parade nights. The task was first to drive 'Typhoon' in a figure of eight in our parade hall without touching the walls. When this basic skill had been mastered we placed some empty Coke cans on the floor and weaved round them without knocking them over.

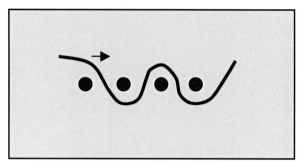

Initial Driver Training Course

There was a huge difference in the level of co-ordination shown by the cadets. Some had great difficulty controlling the robot at all. Others did well when it was travelling away from them, but continually got the controls reversed when it was travelling towards them. For some, the learning curve was steep and they soon became quite skilled at this basic task. This was a difficult exercise and allowed me to choose a short list of 6 drivers (4 boys and 2 girls) to receive extra training.

The advanced training involved moving quickly from one traffic cone to the next and dodging obstacles on the way. To make the task more realistic the weapon cone was given a quick spin (by hand) to ensure the drivers were using the flag to show them which direction the chassis was pointing. Next we set out a number of cans to simulate the edge of the pit and tried driving the robot round this pit at speed. It soon became clear that one young thirteen year old cadet called Gary Cairns, who had only been on the Squadron 6 months, had considerable driving talent and could consistently make Typhoon go where he wanted. Cpl Alistair MacLeod and Cpl Keri Scott also showed good driving skills and temperament.

We knew that driving skill was only one part of the task. Our potential team had to be able to think quickly, formulate tactics and handle an interview without giving monosyllabic answers. I also devised a quiz on aspects of Robot Wars which eliminated those cadets who had never watched the programme. The practice interviews proved very difficult for some cadets who became tongue tied and embarrassed in front of a camera.

Driving practice for the Technogames assault course was very focussed. We again used Coke cans as obstacles and devised a course which tested the basic manoeuvres and skills we thought were needed. The course consisted of a path laid out with several 90 degree corners, a reverse-in garage, some hard obstacles, a sheet of MDF simulating a ramp and a football and goal mouth to finish. This course was extremely difficult to negotiate, and Typhoon Rover was a handful to control, exhibiting a characteristic jerkiness when it turned. Some cadets managed the course in under a minute and Gary was again best with a time of around 40 seconds. We needed a much faster time than that and I decided to try using my test pilot skills to tune the control system.

We first tried exponential control laws for both longitudinal and lateral sticks. Without telling the driver what value was selected I plotted the exponential value against the time to complete the course. The time initially came down and then as the value was increased further the overall time increased again. The exercise was repeated for the other channel and then a final tuning process was undertaken to determine the optimum values. Next we adjusted the maximum power for longitudinal and lateral control to prevent unnecessary skidding. This tuning gave a dramatic improvement in the course time which dropped from 40 to 25 seconds. With more practice we were able to achieve consistent times of around 21 seconds. The biggest noticeable change was that Rover now turned smoothly to the desired angle without overshooting.

The entry form stated we must be able to negotiate a tyre wall and a 1 in 10 ramp. It did not specify how many tyres or how long a ramp. Our local Kwik-Fit garage kindly let us construct a tyre wall in their car park and we were encouraged when Rover was able to force its way through. When we eventually saw the true assault course we were surprised to find it only had one layer of tyres placed vertically side by side and was no barrier to the powerful Rover.

Practice Tyre Wall

The ramp was another unknown as although we knew the angle we did not know whether there would be a small step at the bottom and whether it would have an apex at the top. We designed the ground clearance of Rover for the worst case of a ridge at the top of the ramp, but it turned out to be a straight see-saw with no ridge at all.

93

Black Box Fight Recorder

The black box fight recorder was created by Roger and is used to gather data during testing and real fights. It is an invaluable tool to help us tune our design parameters. The data logger records 16 parameters 5 times every second. It starts recording when the removable link is inserted and always records the last 26 minutes of data until the link is removed.

To calculate the memory required by this recorder we multiply the number of parameters by the resolution (8 bits) x the number of recordings each second x the maximum time of recording:

$$\text{Memory required} = 16 \times 8 \times 5 \times 26 \times 60$$
$$= 1 \text{ million bits}$$

After each test, playback mode is selected by pressing a button on the black box. Data is then downloaded via a serial link into a laptop PC. The data can either be viewed directly in graphical format or manipulated using Microsoft Excel to create graphs of specific parameters of interest. A real time transfer of data via telemetry is available using the 400 MHz band.

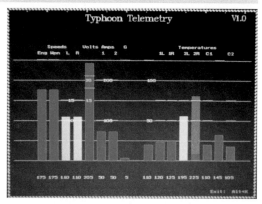

Data Display on Notebook PC

Parameter	Range	Units
Speed of Cone	0 – 1400	rpm
Speed of Engine	0 – 12000	rpm
Speed of Left Rear Wheel Motor	0 – 6000	rpm
Speed of Right Front Wheel Motor	0 – 6000	rpm
Volts of Battery	0 – 30	volts
Amps delivered from Left Battery	0 – 500	Amps
Amps delivered from Right Battery	0 – 500	Amps
Shock (Combined X+Y Horizontal Axis)	0 – 1250	G
Temperature of Ambient Air in Cone	0 – 120	deg C
Temperature of Cable to Removable Link	0 – 120	deg C
Temperature of Left Rear Wheel Motor	0 – 120	deg C
Temperature of Right Front Wheel Motor	0 – 120	deg C
Temperature of Left Speed Controller	0 – 120	deg C
Temperature of Right Speed Controller	0 – 120	deg C
Temperature of Chassis by Exhaust	0 – 120	deg C
Temperature of Engine Crankcase/Clutch	0 – 400	deg C

Sensors

To measure temperature we used thermistors whose resistance changed with temperature. The relationship between temperature and resistance is not linear, so we used a look-up table to calibrate this sensor.

To measure amps we used a commercial transducer which gave a linear output of 0 to 5 volts with increase in current from 0 to 500 amps.

To measure speed we used an infra red transmitter and receiver which detected when the teeth of a plastic disc on the motor shaft were interrupting the light path. The frequency of the pulses was proportional to the speed of rotation.

To measure shock we used piezo-electric shock detectors which are normally fitted to window frames to detect when a burglar breaks the glass. We mounted two detectors at right angles to each other to measure shock (g) in the horizontal X and Y axes. The electrical outputs are combined to give the horizontal g.

We wanted to know how fast the cone was spinning so we decided to fit three high brightness diodes near the top of the cone to indicate three different speeds. To avoid transferring power from the chassis to the rotating cone, which would have needed complicated electrical slip rings, we decided to install a small battery inside the rotating cone. The sensors were simple microswitches mounted to the cone at different radii from the centre.

Thermistor On Motor Speed Controller

Speed Sensor on Wheel Motor

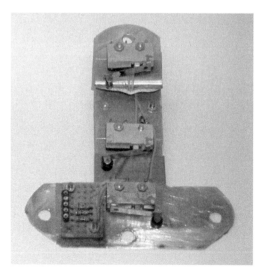

Cone Speed Sensor Microswitches

95

As the speed of rotation increased the outer microswitch closed first as it experienced the greatest centripetal force. This occurred about 300 rpm. The middle one closed next at about 600 rpm and the inner one only closed when the cone reached a near maximum speed of over 900 rpm.

The light emitting diodes were coded to confuse our enemies, but gave valuable tactical information to the team as to how much energy was available to unleash on our opponent.

**High Brightness Light Emitting Diodes
On the Side of the Weapon Cone**

Safety was our greatest concern. We were very conscious that we had a serious destructive weapon which, if spinning, could easily amputate legs. We therefore inhibited the drive to the weapon except when doing specific spin-up trials in the middle of an empty car park.

The radio control failsafes and the removable link contributed to our safety, but a chassis weighing 100kg that can accelerate from 0 to 15 mph in about a second and if spinning is capable of exiting through the brick wall of our parade hut, was a major hazard that must be treated with respect at all times.

Typhoon is a devastating Weapon of Mass Destruction.

Conclusion

The Typhoon Robots Project has stimulated the interest of the cadets in practical engineering, technology and physics and exposed them to education in citizenship, project management and leadership. It has provided an experience and impact on the cadets that they will remember for the rest of their lives. They have experienced major successes in two national televised competitions and this success has been communicated within their schools, to the local community and to the national air cadet community.

By fighting in a war every year 870 (Dreghorn) Squadron is truly a front line operational air cadet unit.

Appendix A
Our Robot Sponsors

Sponsor	Product
Aardvark Electronic Components	Safety Break Connectors
Aerospace Machine Technology (AMT)	Welding
Apollo Logistics	Titanium Sheet
Arnold Clark	Cash
BAE Systems	Cash
BAE Systems Plymouth	Silicon Gyro
BAE Systems Samlesbury HTTF	Machining of Weapon Ring and Welding of Titanium Cone
Coraza Systems Europe	Aluminium Cones
D Moffat (Jewel & Esk Valley College)	Welding
Edinburgh and South Scotland Wing ATC	Cash
Eurofighter GmbH	Cash
European Aeronautics Defence and Space	Cash
Hexcel	Aluminium Honeycomb
Hunters Scrap Yard	Weapon Ring
Hypertac	Safety Break Connectors
JL Steel Services	Hardox Claws
Kwik-Fit	Tyre Wall
Loctite UK Ltd	Glue
Master Cameron Miller	Child's Bicycle
MG Rover	Cash
Mr Brian Armistead	Metal
Reserve Forces & Cadet Association	Cash
SAFT UK	NiMH Batteries
Taylor's Garage	Welding
TRIAD AS (Norway)	Cash
W L Gore	Team Jackets & Gortex Skirt
1936 (Newton) Squadron ATC	Accommodation During Filming
Our long suffering Wives	Sustenance

Appendix B
The Typhoon Family

Typhoon

Typhoon was our first robot and the one that proved our concept and design. We have made many modifications over the last 3 years, but the cone remains the same and shows evidence of its numerous conquests.

'Typhoon' (50 kg)

Pilot:	Cpl Gary Cairns,	
WSOs:	Cadet Hazel Taylor	2001
	Sgt Graeme Horne	2002
	Cadet Tina Bowman	2003

Specification	
Weight:	50 kg (Middleweight class)
Power:	Three 24 v Electric Motors (3.2 HP)
Control:	40 MHz Six channel PCM radio
	Silicon Gyro
Weapon:	High inertia steel rotating ring
	2 primary cutters and 2 ground skimming cutters
Rotating Mass:	23 kg
Cutter Diameter:	0.9 m
Cutter Speed:	>100 mph
Kinetic Energy:	>18 Kilo Joules
Self Righting:	N/A. (Due to Gyroscopic properties of weapon ring)
Countermeasures:	Robot Secret

'870' (Goalkeeper)

'870' is actually our Robot Wars middleweight champion fighting robot 'Typhoon' with its weapon cone inhibited and fixed. It is a dedicated goalkeeper and has large white hands fixed to its front and back. Because it had previously appeared in Robot Wars it could not be called 'Typhoon' and had to be painted in a different colour scheme.

870

Captain:	Sergeant Graeme Horne
Pilot:	Corporal Andrew Fullerton
Tactician:	Cadet Naomi McGeary

Specification

Weight:	52 kg
Length:	1.6m (Maximum length allowed is 2m)
Power:	Two 24 v Electric Motors (Producing 2 HP)
Control:	40 MHz Six channel PCM radio
	Silicon Gyro

Typhoon Rover

'Typhoon Rover' was specially designed for the assault course with its wooden chassis optimised to go up ramps and steer a football into a goal. Typhoon Rover contains the main parts of our heavyweight fighting robot 'Typhoon 2' and is powered by four 1 Horse Power electric motors with independent drive to each of its four wheels.

Typhoon Rover

Pilot: Cadet Gary Cairns
Tactician: Cadet Jonna Salvesen   (Assault Course)
                  Cadet Andrew Snedden (Football)

Specification
Weight: 70 kg
Power: Four 24 v Electric Motors (Producing 4.2 HP)
Control: 40 MHz Six channel PCM radio
Construction: Wood Chassis and Aluminum top

Byphoon

Byphoon is a modified child's bicycle. It is designed to be self balancing and to power itself around a flat oval circuit, approximately 50 metres in length. Stability is achieved by a large instructional gyroscope fitted to the frame and a closed loop stabilisation system using a pendulum. This was connected to a potentiometer which drove two PWM servos connected to the handlebars.

To prevent the pendulum oscillating a damping mechanism was fitted which used washing-up liquid in an aluminium pot.

To achieve a controlled turn we mounted the battery (a heavy mass) on a screwed rod which was mounted laterally and driven by an electric drill motor. Moving the battery in the direction we wanted to turn allowed the C of G to be moved sideways rapidly. The radius of turn was proportional to the lateral movement of the C of G and the stability was achieved by the closed loop steering system that steered in the direction of the lean until the centripetal force balanced the gravitational force and returned the pendulum to a near central position of equilibrium.

Byphoon

Driver:	Cadet Christopher MacPherson
Tactician:	Cadet Edward Graham

Specification	
Mass	15kg
Length	0.9m
Height	0.5m
Width	0.5m
Power	Two 200W 18 volt drill motors
Control:	40 MHz Six channel PCM radio

Typhoon 2

The Typhoon 2 robot engineering and systems have some similarities to Eurofighter. The chassis is made from aluminium honeycomb; the outer cone armour is fabricated from titanium sheet and separate electrical systems are incorporated to minimise single point failures following battle damage.

The heavy steel outer ring and cone is powered by a chain saw petrol engine and rotates at up to 1000 RPM with the weapon cutters travelling at over 110 mph. The cutters are made of Hardox, an extremely hard Swedish steel that is normally used for digger buckets. The weapon cone acts as a huge gyroscope so no self-righting mechanism is fitted.

The most unusual component is the high tech device designed to eliminate wobble of the cone at high speed. This is called the 'Wobulator' and its design is a close kept military secret.

Heavyweight 'Typhoon 2' (100 kg)

Pilot:	Cpl Gary Cairns,
WSO:	Sgt Graeme Horne

Specification	
Weight:	100 kg (Heavyweight class)
Power:	Four 24 v Electric Motors (Producing 4.2 HP)
	2 stroke petrol engine (Producing 5 HP)
Control:	40 MHz Six channel PCM radio
Weapon:	High inertia steel rotating ring
	2 primary cutters & 2 ground skimming cutters
Top Armour	Titanium
Rotating Mass:	48 kg
Cutter Diameter:	1.0 m
Cutter Speed:	>110 mph
Kinetic Energy:	>39 Kilo Joules

Typhoon Thunder and Typhoon Lightning

These lightweight versions of Typhoon weighed 25kg each. The bases are made out of MDF and the webs are aluminium honeycomb. We use 18 volt electric drill motors from B & Q to power these robots; two for the wheels and two for the weapon. These cordless drills not only provide a motor and gearbox, but also a NiCad battery pack and a battery charger. The girls made a pleated skirt for their robot 'Thunder' and this was velcroed to the cone.

Typhoon Thunder (25kg)
  Built by female cadets
  Pilot: Cpl Keri Scott
  WSO Cadet Amy Drinkwater

Typhoon Lightning (25kg)
  Built by male cadets
  Pilot: Cpl Alistair MacLeod

Specification	
Weight:	25 kg (Lightweight class)
Power:	Four 18 v Electric Drill Motors
Control:	40 MHz Six channel PCM radio
Weapon:	High inertia steel rotating ring (fabricated)
	2 primary cutters
	2 ground skimming cutters
Rotating Mass:	12 kg
Cutter Diameter:	0.7 m
Cutter Speed:	>100 mph
Kinetic Energy:	>5 Kilo Joules
Self Righting:	N/A. (Due to Gyroscopic properties of weapon ring)
Countermeasures:	Robot Secret

Typhoon Twins

Thunder and Lightning were teamed up as a clusterbot for the middleweight competition. The two cones were joined together with light wire tied between the claws. To make the twins into a single unit for entry into the arena the cones were joined with paper. This came off easily when the two separated at the start.

For the 2002 Extreme 2 the rules were changed so that more than 50% of the clusterbot had to be immobilised to eliminate it. The rule was changed back the following year to 50% or more immobilised. This was a pity and does little to encourage clusterbots which are liked by the audience.

Typhoon Twins Clusterbot (50kg)
Pilots: Cpl Keri Scott & Cpl Alistair MacLeod

Specification	
Weight:	25 kg (Lightweight class)
Power:	Four 18 v Electric Drill Motors
Control:	40 MHz Four channel PCM radio
Weapon:	High inertia steel rotating ring (fabricated)
	2 primary cutters
	2 ground skimming cutters
Rotating Mass:	12 kg
Cutter Diameter:	0.7 m
Cutter Speed:	>100 mph
Kinetic Energy:	>5 Kilo Joules
Self Righting:	N/A. (Due to Gyroscopic properties of weapon ring)
Countermeasures:	Robot Secret

Typhoon Cadet

Typhoon Cadet is a 12kg Featherweight and the newest addition to our family. It uses three 18 volt electric drill motors; Two for the wheels and one for the weapon. It provided us with a huge challenge to get the weight below the limit and we had to drill lightening holes in every piece of metal.

Typhoon Cadet (12kg)

Pilot: Cpl Gary Cairns
WSO: Cadet Amy Drinkwater

Specification	
Weight:	12 kg (Featherweight class)
Power:	Three 18v Electric Drill Motors
Control:	40 MHz Six channel PCM radio
Weapon:	High inertia steel rotating ring (fabricated)
	2 primary cutters
	2 ground skimming cutters
Rotating Mass:	5 kg
Cutter Diameter:	0.6 m
Cutter Speed:	Calculate yourself (Appendix C)
Kinetic Energy:	Calculate yourself (Appendix C)
Self Righting:	N/A. (Due to Gyroscopic properties of weapon ring)
Countermeasures:	Robot Secret

Appendix C
Design Your Own Featherweight Typhoon

Calculate the parameters of Typhoon Cadet yourself and try modifying some of these parameters to improve or personalise your full body spinner featherweight robot.
How do your modifications change other parameters?
How will your modifications effect the fighting ability?
Remember that any modification must not increase the total mass above 12kg.
All the equations and methods are explained in the previous chapters.

The relevant design data for the current Typhoon Cadet is:

The maximum no-load speed of each drill motor	20400 rpm.
The drill motors are fitted with epicyclic gearboxes with gear ratio of	24 : 1
The wheels are fixed directly onto the output shaft of the gearbox	
Wheel radius	0.05m.
To Convert from metres /sec to mph multiply by	2.237
Stall Torque of wheel motors	0.323 Nm.
Number of wheels	2
Mass of Typhoon Cadet (Featherweight Class)	12 kg
Coefficient of friction	0.9
Stall Torque of weapon motor	0.323 Nm
Weapon motor to cone gearing = 24:1 and 1.2 : 1	28.8 : 1
Distance of wheels from centre	0.12m
Radius of cutters	0.3m
Weapon Ring External Radius	0.225m
Thickness of weapon ring	6mm
Height of weapon ring	0.05m
Equivalent thickness of aluminium cone	2mm
Mass of each claw/cutter	0.45 kg
Number of claws	2
Radius of C of G of cutters	0.26m
Density of steel	7850 kg/m^3
Density of aluminum	2700 kg/m^3
Battery voltage	16.8 volts
Battery capacity with high current drain	3.0 Ah
Internal Battery Resistance	0.042 ohms
Motor Power	172 Watts
Motor Stall Current	52 Amps
Motor Internal Resistance	0.3 ohms
Average amps used by wheel motor	10 A
Average amps used by weapon motor	15 A

Calculate:

1.	The maximum wheel speed	
2.	The maximum horizontal speed.	
3.	Maximum wheel torque	
4.	Maximum wheel force on ground from 1 motor	
5.	Limiting frictional force from 2 wheels	
6.	Limiting frictional force from 1 wheel	
7.	Wheel speed for limiting frictional force	
8.	Longitudinal speed for limiting frictional force	
9.	Longitudinal acceleration	
10.	Time to accelerate to 6.5 mph	
11.	Distance travelled to 6.5 mph	
12.	Kinetic Energy of the Typhoon Cadet Chassis	
13.	Weapon torque	
14.	Rotational force (steering bias) at wheels due to weapon	
15.	Mass of ring	
16.	Maximum Speed of Weapon cone Rotation	
17.	Peripheral speed of cutters	
18.	M of I of ring	
19.	M of I of Cone	
20.	M of I of Claws	
21.	Total M of I of weapon	
22.	Maximum weapon energy	
23.	Height energy could theoretically toss another 12 kg robot	
24.	Time to spin up to 700 rpm assuming motor efficiency 65%	
25.	Battery capacity required for 5 minute fight	
26.	Volts dropped across battery	
27.	Heat generated in battery	

Answers to Annex C calculations: 1) 850 rpm or 14.7rps. 2) 4.45mps or 9.96 mph. 3) 7.75 Nm. 4) 155 N. 5) 108 N. 6) 54 N. 7) 554 rpm. 8) 6.5 mph or 2.9 mps. 9) 9 m/s^2. 10) 0.32 sec. 11) 0.46 m. 12) 119 Joules. 13) 9.3 Nm. 14) 77.5 N per wheel. 15) 3.285 kg. 16) 708 rpm or 74 rad/sec. 17) 42.2 m/s or 49.7 mph. 18) 0.162 Kgm^2. 19) 0.022 kgm^2. 20) 0.055 kgm^2. 21) 0.239 kgm^2. 22) 657 Joules. 23) 5.475m. 24) 5.9 sec. 25) 2.9 Ah. 26) 1.47 volts. 27) 51Watts.

Appendix D
Physical Quantities

Linear Quantity	Sym -bol	Units	Equation	Rotary Quantity	Sym -bol	Units	Equation
Mass	m	kg		Moment of Inertia	I	kgm^2	Several
Force	F	N	F=ma	Torque	T	Nm	$I\omega$
Distance	d	m		Angle	θ	radians (rad) revolutions	2π rad =1 rev
Velocity	v	m/s	v=d/t	Angular Velocity	ω	rad/s or rpm	
Acceleration	a	m/s^2	a=v/t	Angular Acceleration	α	rad/s^2	
Potential Energy	E	joules	E=mgh				
Kinetic Energy	E	joules	$E=\frac{1}{2}mv^2$	Kinetic Energy	E	joules	$\frac{1}{2}I\omega^2$
Electrical Energy	E	Joules	VIt				
Mechanical Power	P	Watts or HP	P=Fv 1 W=1 J/s P=mav 1HP=746W	Mechanical Power	P		$P=T\omega$
Electrical Power	P	Watts	VI	Electrical Power	P		$P=T\omega$
Momentum	M_L	kgm/s	mv	Angular Momentum	M_A	kgm^2/s	$M_A=I\omega$
Density	ρ	kg/m^3					
Battery Capacity		Ah					
Coefficient of Friction	μ	None					

A **Newton** is the **Force** required to accelerate a mass of 1 kilogram at a rate of 1 metre per second per second.

A **Mass** of 1 kg on earth has a **Weight** of 10N. (Acceleration due to gravity is 10 m/s^2)

Work done is the application of a force over a distance. (measured in Joules)

A **Joule** is a force of 1 Newton acting through a distance of 1 metre (1Nm)

Energy is the capability of doing work (measured in Joules)

Power is a measure of how quickly work can be done. i.e. How quickly you can apply force over a distance

A Power of 1 **Watt** is a force of 1 Newton applied at a speed of 1 metre per second.

1 Watt = 1 Newton-metre per second or 1 Joule per second

Momentum is the product of mass and velocity

Index

7th Wars, 28, 58, 68, 77, 90

870 Squadron, 3, 7, 10, 14, 17, 21, 23, 24, 26, 35, 43, 91, 95, 112

Acceleration, 48, 51, 52, 59, 70, 109
  angular, 59, 109
  longitudinal, 29, 51

Aerial, 85

Air Cadets, 4, 7, 8, 11, 20, 21

Air Training Corps, 21

Aluminium, 40, 42, 65, 97
  honeycomb, 97

Angle, 59, 109

Armour, 11, 18, 29, 37, 41, 42, 60, 65, 102

Assault Course, 22, 54, 100

Axe, 9, 58

Back emf, 55

BAE Systems, 4, 10, 12, 36, 43, 89, 113

Balancing, 81

Battering Rams, 9

Battery, 75, 79, 93, 97, 106
  capacity, 14, 75, 106, 107
  charger, 81
  Nickel Cadmium (NiCad), 79
  Nickel Metal Hydride (NiMH), 16, 79, 80
  Sealed Lead Acid (SLA), 79, 80
  technology, 79

Bicycle, 97

Brain-Storming, 9, 10

Byphoon, 4, 24, 101

CAD (Computer Aided Design), 43

Chain Strength, 52

Claws, 65, 97, 107

Clusterbot, 104

Clutch, 67, 68, 78, 93

CO_2, 9

Construction, 15, 40, 44

Control Laws, 89

Current, 54, 106

Cutter Bite, 61, 63

Cycling, 22, 24

Damage, 10, 45, 58

Density, 61, 65, 106, 109

Destruction, 69

Disciplines, 12

Discs, 9, 58

Distance Travelled, 52

Drive ratio, 43, 46, 47, 49

Driver Training, 91

Edinburgh Science Festival, 27

Electric Motor, 49, 84
  speed controller, 16, 54, 85, 86
  wheel, 54

Electrical Components and Connections, 81

Energy, 10, 54, 58, 59, 69, 70, 109
  chassis, 59
  kinetic, 58, 59, 107, 109
  replacing, 69
  rotary, 10
  weapon, 59

Engineering, 13

Eurofighter, 11, 14, 16, 26, 29, 34, 97, 102

Extreme-2, 26, 38, 104

Failsafe, 84

Flag, 11, 14, 18, 27, 91

Flipper, 9

Flywheel, 10

Football, 23, 54, 100

Force, 48, 51, 59, 62, 70, 109
  centripetal, 62, 63, 67, 94, 101
  linear, 48, 70
  rotary, 48

Freewheel, 77

Frequency Modulation, 84

Friction, 50, 109
  coefficient, 50, 106
  limiting frictional force, 50, 53

Fuel Capacity, 78

G, 43, 70, 93

Grand Champion, 8, 112

Gravity, 39, 58

Gyro, 97

Gyroscope, 10, 11, 29, 38

Hardox, 29, 41, 65, 97, 102

Harrier, 11

Heat, 55, 58, 86, 107

Honours, 36

Horse Power, 22, 45, 46, 53, 69, 78, 100

House Robots, 7, 37, 69

Hypno-Disc, 9, 18, 58, 66

Impact, 70

Impulse, 70

KISS (Keep It Simple, Stupid), 17

Mass imbalance, 61, 62

Materials, 40

Matilda, 20
MDF (Medium Density Fibreboard), 41, 92, 103
Medal, 22, 24
Moment of Inertia, 59, 60, 63, 64, 65, 109
Momentum, 70, 109
MOSFETS, 86, 89
Petrol engine, 66, 67, 77, 78, 90
Physical Quantities, 109
Pincers, 9
Pneumatics, 9
Power, 53, 70, 75, 77, 83, 106, 109
    electrical, 53, 54, 109
    mechanical, 53, 54, 109
    required by weapon, 76
    required by wheels, 75
Publicity, 12, 20, 21
Pulse Code Modulation, 84
Pulse Width Modulated, 84
Radar Engineers, 4
Radio Control, 84
RAF, 7, 8, 13, 17, 21, 23, 24, 26, 34
Receiver, 84
Recorder, 13, 14, 93
Ref Bot, 7
Reliability, 14
Removable Link, 11, 82, 83, 93
Resistance, 75, 82, 106
Robot Wars, 3, 4, 7, 8, 9, 10, 17, 23, 26, 28, 36, 77, 91
Safety, 14, 95, 97
Scream Team, 30
Self-Righting, 11, 38
Sensors, 94
Servos, 16, 24, 77, 101
Shock, 43, 70, 93
Shock Absorbers, 43
Skills, 12
Specific gravity, 79
Speed
    cutters, 64
    maximum horizontal, 47
    no wheel slip, 51
    rotation, 61
Speed controllers, 85
Spinning Horizontal Drums, 9
Sponsors, 20, 97
Steel, 41, 65, 97
Steering Bias, 56

Team Selection, 17
Technogames, 3, 4, 17, 22, 23, 24, 36, 54, 89, 92, 112
Tests, 15
Time to Spin Up, 61, 66
Titanium, 40, 42, 65, 97
Torque, 48, 49, 55, 57, 59, 67, 78, 106, 109
    weapon, 56
Transmitter, 90
Trophy, 8, 112
Typhoon, 11, 13, 14, 15, 16, 17, 20, 21, 22, 23, 24, 26, 29, 35, 37, 38, 39, 40, 41, 42, 43, 44, 45, 47, 49, 56, 57, 58, 59, 60, 62, 64, 68, 69, 70, 74, 79, 81, 82, 85, 89, 90, 91, 98, 99, 103, 106, 112
Typhoon 870, 23, 24, 99
Typhoon Cadet, 29, 105, 106, 107
Typhoon Family, 10, 26, 45, 98
Typhoon Lightning, 13, 21, 26, 27, 44, 103, 104
Typhoon Robots Project, 4, 12, 95
Typhoon Rover, 22, 23, 54, 79, 89, 92, 97, 100, 112
Typhoon Thunder, 21, 26, 27, 103, 104
Typhoon Twins, 13, 26, 27, 104
Typhoon-2, 7, 8, 21, 22, 26, 28, 29, 30, 34, 35, 37, 41, 42, 43, 45, 46, 50, 57, 58, 60, 65, 66, 67, 68, 69, 70, 74, 82, 86, 89, 90, 100, 102
Tyre Wall, 22, 92, 97
Velocity, 51, 59, 109
    angular, 59, 109
Volts Dropped, 107
Weapon, 56, 57, 65, 76, 90, 106, 107
    cone, 43, 65, 66, 93, 95, 97, 106, 107
    control box, 16, 90
Weapon of Mass Destruction, 58
Weapon Ring, 11, 15, 41, 43, 60, 65, 74, 97, mass, 60
    size, 60
Weapon Systems Operator, 16, 17, 34, 90
Wedges, 9
Weight, 76, 109
Wheel Arrangement, 45
Wiring, 81
Wobulator, 30, 102
World Record, 22

About The Author

Peter Bennett spent 21 years in the Royal Air Force and flew the Harrier Jump Jet operationally with No 4 Squadron and No 1(F) Squadron. He graduated from the Empire Test Pilot's School in 1977 and was a Test Pilot on the Harrier, BAC 1-11 airliner and Wessex helicopter at the Royal Aircraft Establishment Bedford. After managing the Nightbird Harrier research programme at Farnborough he retired from the RAF and joined the Avionics industry. He currently works for BAE Systems as their Business Development Director – Euroradar. In his spare time he is an RAF volunteer reserve officer with No 870 Squadron of the Air Training Corps based in Edinburgh, Scotland. He teaches aeronautics, organises the cadets powered flying and gliding and manages the Typhoon Robots Project.

There are 928 Air Cadet Squadrons throughout the UK.

Minimum age to join is 13 years

www.aircadets.org

www.typhoon-robot.co.uk

Acknowledgements

James Wilson	Spare boxer shorts article
Mentorn	Picture of Rover approaching ramp in Technogames
Giles Barnard	Picture of team with Grand Champion Trophy
The Typhoon Team	The Cadets of 870 (Dreghorn) Squadron and their Instructors from BAE Systems Avionics
Our Long Suffering Wives	